T5-DGI-451

Designing Knowledge-Based Systems

Designing Knowledge-Based Systems

PRENTICE-HALL
Englewood Cliffs, New Jersey 07632

First published in 1985 by Kogan Page Ltd
120 Pentonville Road, London N1 9JN

This edition published 1986 by
Prentice-Hall
A Division of Simon & Schuster, Inc.
Englewood Cliffs, New Jersey 07632
ISBN (Prentice-Hall edition) 0-13-201823-3

Printed in Great Britain

To Pam, Elizabeth and David

> For myself, I am interested in science and in philosophy only because I want to learn something about the riddle of the world in which we live, and the riddle of man's knowledge of that world.
>
> *K.R. Popper (1959)*
> *The Logic of Scientific Discovery*

Contents

Preface

Chapter 1: Knowledge for the New Generation Computers 13

Introduction 13
An approach to knowledge 14
Applied epistemics 19
The knowledge base 20
Knowledge engineering 23

Chapter 2: Relational Analysis 27

Introduction 27
Relations and tables 30
Functional dependency 32
Two important relational operations 33
Normalization and the computer file model 35
Normalization and the relational model 39
The role of relational analysis 45

Chapter 3: The Conceptual Model 47

Introduction 47
Extended Relational Analysis 51
Representations of a conceptual model 58
The effects of multivalued and join dependencies on ERA 62
ERA as an elicitation technique 68
Conclusions 78

Chapter 4: Engineering Conceptual Databases 79

Introduction 79
Logical models and schemas 81
Storage structures on the Content Addressable File Store (CAFS) 97
The choice 110

Chapter 5: A Conceptual Database Interpreter 113

Introduction 113
System outline 114
The mechanism of implicit queries 115
Mechanics of insertion 119
Examples of insertion 121
Examples of deletion 125

Interaction of language and the conceptual model 128
The effectiveness of the conceptual model 128
Limitations of ERA 129

Chapter 6: A Conceptual Database Consultant System 133

Introduction 133
The CRIB system 135
CRIB on CAFS 140
A simplified CRIB system 147
Conclusions 149

Chapter 7: The Structure of Artificial Intelligence Programs 151

Introduction 151
Production systems 154
Graph search 167
Decomposable production systems 174

Chapter 8: Mechanical Reasoning 181

Introduction 181
A calculus of reason 182
Reasoning with predicates 193
Rule-based systems 204
The limits of machine reason 217

Chapter 9: Applied Epistemics 223

Introduction 223
Uncertain knowledge 229
MYCIN 242
Types of epistemic systems 247
Knowledge refining 249
Conclusions 262

Chapter 10: Computer 'Understanding' 265

Introduction 265
First generation systems 267
Second generation systems 268
The introduction of semantics 270
Case grammar: the limited universe 276
Word-government: the syntactic/semantic boundary 278
Semantic structures: memory and meaning 281
Frames and scripts 286
Discussion 288
Conclusions 292

Epilogue 295

Appendix I: the definition of application DEMOB 297

Appendix II: BNF (MOD) definition of three states of FIDL 299

References and Bibliography 303

Author Index 311

Subject Index 313

Preface

This book concerns a design practice for knowledge-based systems, sometimes referred to as Intelligent Knowledge-Based Systems (IKBS). It is a practice that has firm foundations in both experience and theory. Such a design practice will ensure that knowledge-based systems are created functionally correct by conforming to our perception of the world, are made robust by using well-established principles, and are formed for simplicity of control by blending with our modes of expression. The support for good design necessarily comes from an understanding at sufficient depth of a wide range of related subjects.

Designing a knowledge-based system that uses a very large number of facts requires techniques that are drawn from database theory, computer science and artificial intelligence. Database theory has evolved to cope with the design of mass storage structures for multiple applications. Artificial intelligence has been concerned with the complex relationships of facts stored in main memory. The issues within each of the fields are strikingly different. Yet both fields share the use of the computer for their realization. This book describes a method for managing the design of large scale database systems for problems which may involve techniques derived from artificial intelligence.

The complete book is primarily intended for computer science graduates and undergraduates who have specialized in either database theory or artificial intelligence and who would like to have an understanding of the relationships between these fields. In order that a comparison can be made between the two fields, many familiar examples have been analysed using a common unifying technique.

This book is also a personal account of the insights I have

gained through attempting to answer the question, 'What could it mean to design a machine that understands?'. In this attempt I have met many people who have helped me to clarify issues. Some of them have sketched in outline broad exciting territories, others have described with absolute precision small but important areas, all have been friends and colleagues.

Chapter 1 contains material drawn from the paper *Knowledge for Machines* (Addis and Johnson, 1983). The theory of knowledge adopted in this paper is mainly attributable to Leslie Johnson. This paper reflects a change in attitude about the nature of knowledge; it is a change that moves the view of knowledge away from being states of the brain to the view that knowledge is a shared human response to a complex environment. At one level this change has little effect on the knowledge representation schemes implemented in artificial intelligence programs. What does alter with this change is the interpretation of what these programs achieve. It describes clearly the role of a knowledge-based system. Chapter 1 does not explore other views of knowledge since even an adequate comparison would be very extensive. However, references are given to the literature that will allow the interested reader to follow alternative views.

Chapters 2 and 3 give a reassessment of my understanding of relational analysis and Extended Relational Analysis (ERA). My original understanding was formed during the initial development stages of the ICL Content Addressable File Store (CAFS). However, in the light of recent work on relational database theory, many of my original concepts have proved to be incomplete. ERA is a method of expression which helps in the determination of the components and the structure of complex environments. The method is still evolving and a precise specification language, independent of computer architecture, is expected to emerge.

Chapter 4 describes an implementation stage that is often ignored because it lies on the boundary between software and hardware. Eventually each system will have its definition embodied in a machine, and that machine will have properties that influence performance. Peter King (Birkbeck College, University of London) in the early 1970s suggested to Victor Maller, Edward Babb and myself (then all employed at ICL RADC) the links between relational analysis and various implementation models. The development of the CAFS engine during this time was influenced by many people, and it would be pointless to try to untangle the web of interacting ideas that helped define the CAFS. Among those who influenced the formation of CAFS was George Coulouris (Queen Mary College, University of London). However, the father of CAFS is the late Roy Mitchell who, as my manager, gave me the freedom to think and explore. Such support is a rare gift and was made possible through Gordon Scarrott (Manager ICL RADC).

Chapters 5 and 6 describe two of the general systems that have evolved from the work on CAFS and ERA. Roger Hartley and Frank George did much of the preliminary work on the Computer Retrieval Incidence Bank (CRIB) system before an ERA was done. This alternative approach to diagnostic consultation systems stemmed from a style conceived in conjunction with the users. Chapters 2, 3, 4 and 5 cover approximately two-thirds of the undergraduate module *Formal Aspects of Systems Analysis* at Brunel University.

Chapters 7 and 8 are based upon part of the module *Techniques of Artificial Intelligence* which is given to the fourth year Computer Science undergraduates and to the MSc Computing Systems students. The main course book for this module is *Principles of Artificial Intelligence* by Nils J Nilsson (1980). This was the first book on artificial intelligence I found that covered in sufficient depth the theory behind rule-based systems without the excesses of a complete formal description. However, I felt a need to expand some explanations and to omit others. The two chapters 7 and 8 should be considered as complementary to the first six chapters of Nilsson's book. Chris Reade (Brunel University) made several suggestions concerning the content of Chapter 8, some of which have been incorporated.

Chapter 9 reviews some of the techniques of expert systems in depth, and Chapter 10 is an updated survey of computer 'understanding'. This last chapter draws upon the paper *Machine Understanding of Natural Language* (Addis, 1977) which was triggered by a short intensive period at the Institute for Semantic and Cognitive Studies in Switzerland during 1975. It was here that Yorick Wilks indicated to me the important role of philosophy in technology. These last two chapters form a further part of the module *Techniques of Artificial Intelligence.*

I should like to acknowledge the comments and suggestions made by Max Bramer (Thames Polytechnic), Hamish Carmichael (ICL), Leslie Johnson (Brunel University) and Ladislav Kohout (Brunel University). Time to do my research, write and now teach has been achieved only through the total support of my wife Pam and the patience of my children. My thanks also to Pam Denham, Barbara Yates and Karen Thurley who typed the manuscript.

T. R. Addis
Department of Computer Science
Brunel University
June 1985

CHAPTER 1

Knowledge for the New Generation Computers

> We river-bankers, we hardly ever come here by ourselves. If we have to come, we come in couples, at least; then we're generally all right. Besides, there are a hundred things one has to know, which we understand all about and you don't as yet. I mean pass-words, and signs, and sayings which have power and effect, and plants you carry in your pocket, and verses you repeat, and dodges and tricks you practise; all simple enough when you know them, but they've got to be known if you're small, or you'll find yourself in trouble.
>
> Kenneth Grahame (1908)
> *The Wind in the Willows*

Introduction

Innovation can often be accounted for by a reassessment of old ideas. The development of succeeding generations of computers is marked by new views of current activities, and these new views encourage extensions to the techniques employed. Sometimes these new views come well before the technology is capable of supporting them, and, consequently, the views remain in the backwaters of mainstream science ready to be rediscovered. For example the ideas of Charles Babbage (1792-1871) outstripped the technological resources of the time. Similarly today, the aspirations of those involved in artificial intelligence are only just beginning to be supported. Industry is now interested in harnessing new technology by drawing upon its potential for intelligent behaviour thus giving commercial significance to artificial intelligence. However, it should be made clear that the motives of the people pursuing artificial intelligence are different from those of the people concerned with commercial computing.

On close examination, there will be recognized two major

approaches to artificial intelligence: that of the 'hard' school whose members (sometimes referred to as the 'neats') are concerned with building a strong theoretical component to their work based on pure mathematics; and that of the 'soft' school whose members (the 'scruffies') consider that the strong theoretical component is not only unnecessary but positively harmful (Bundy, 1982). It is the hard school which has influenced the ideas of the new generation of computers. Hence, there is no commitment by the designers of the new generation of computers to purse the more general aims of artificial intelligence, only the strong theoretical component. The important result of artificial intelligence that interests these designers is that a higher level of problem specification can be achieved by engineering 'knowledge' (Newell, 1980; Stefik and Conway, 1982), and it is this principle that underlies the new design objectives. Nevertheless, this hard approach brings with it overtones from the soft school whose pursuits engender a far richer interpretation of machine behaviour than can strictly be justified. It is these overtones that cause high expectations of artificial intelligence.

The new generation computers are hence characterized by the change in view that 'knowledge' rather than 'data' is the essential raw material to be processed. This insight has convinced many people that the current difficulties in program specification, man-computer communication and system development will be dramatically reduced by building machines based on knowledge. Yet what will change? The problems of modern computing can usually be traced to the frustrations experienced in dealing with machines that are perceived to have many features of the human intellect and yet lack understanding. Hence, it was reasoned that machine 'understanding' would reduce our problems. This 'understanding' could be achieved by giving the computer 'knowledge'. There is a danger that words like 'knowledge', 'understanding' and 'meaning', which are endowed with deep human connotations, may create the illusion that machines have somehow transcended their primary properties. One aspect of this book is to explore the extent to which this transcendency is possible.

An approach to knowledge

Knowledge, meaning and understanding are ideas that have usually been examined within the human framework.

'Knowledge', within this framework, is considered to be 'justified true belief', and a difficulty then arises in providing criteria that establish certainty (ie what could be true). From the information technologist's point of view, this concern with the criteria of certainty is irrelevant; the technologist is prepared to leave such problems to the users of his knowledge-based systems. However, the technologist is concerned with the parameters that govern the use of a 'knowledge base' since the purpose of his systems is to provide meaningful, understandable and acceptable responses to users. The responses must not distort the 'knowledge' or introduce unacceptable inferences from man-computer dialogues, but they must present and manipulate 'knowledge' in a way that retains acceptable consistency. Much of the work described in this book will be concerned with the different kinds of consistency that need to be maintained. Note that 'knowledge' (in quotes) is what the user may consider as knowledge; the technologist need have no opinion.

MEANING AND UNDERSTANDING

The technologist needs to be able to distinguish between acceptable or unacceptable 'conversations'. Acceptable conversations are ones which conform to the user's habits of expression and inferential behaviour. These conversations should provide fruitful interactions that are linked with the purposes of the system. There are thus three areas of concern: the language employed, the mode of reasoning used and the range of tasks performed by the user.

It is easily shown that the generation of statements according to the 'syntax' of the signs of the language does not in itself provide meaningful conversations. The sentence 'Green ideas sleep furiously' is an illustration of a grammatical but meaningless statement. However, contexts have been contrived to give such a sentence meaning, and it might be concluded that *any* string of signs can be meaningful. However, the rules by which assignment of meaning to statements is achieved are not clear from an examination of the sentences alone: contexts need to be invented, and stories created, to make a character string meaningful.

Visual display units (VDU) or other outputs by a computer cannot by themselves make meaningful signs. What is needed is a context of shared assumptions, situations and subcultures for a string to have meaning. It becomes necessary (Johnson,

1983) to deny the once prevalent view of meaning that there is in a language an intrinsic connection between signs and the world. This once prevalent view evolved from a description of meaning in terms of reference. Thus, the meaning of a word was the object or relationship in the world to which it was linked. This link to the world of the symbols in the language endowed them with important properties, one of which was the potential of supporting knowledge. It is not surprising that this view considered symbol manipulation as the route to meaningful intelligence. It is a view that is still retained by many artificial intelligence workers (Simon, 1969); it is a view that is difficult to abandon due to its central role in the technological culture. However, the view is flawed because it demands absolute properties of language that are independent of human culture. Modern philosophical thinking has been forced to consider alternative approaches to account for meaning since no relevant absolute properties could be fully justified. One of these approaches considers that language and meaning can only be explained in terms of socially agreed 'understanding' of words. This understanding is derived from the procedures of justification that are supported by accepted (but not absolute) primitive concepts. It will be found that it is these primitive concepts that are employed by those interested in artificial intelligence to create their knowledge bases. (For a description of the once prevalent view see Harrison, 1979.)

THE KNOWLEDGE LEVEL

Historically, the development of knowledge representation schemes has been associated with the representation of meaning. This association stems from the deeply held underlying theory about the nature of knowledge which does not contain the social element. From the perspective presented here, the relationship between meaning and knowledge is a subtle one; but this is not an issue that will be explored too deeply. There are, however, levels of abstraction associated with knowledge. A knowledge structure need not have any one-to-one correspondence with structures that sustain meaning, but it is important that it captures the correct level of abstraction. Knowledge elicitation techniques that extract 'knowledge' from the users of systems and the experts employed to provide the knowledge are designed to identify this structure. The method of Extended Relational Analysis (ERA) described in Chapter 3 is one way of elucidating aspects of such structures. The meaning of this structure need not be represented. It is this meaning of the structure and the symbols employed that lies outside the domain of the machine. It must be emphasized, however, that it is not being asserted that the knowledge level and meanings involved are completely independent but only that they can be treated as independent for our purposes. The users know how to understand language, and this capacity can be drawn upon to understand the computer output derived from the internal structures. It is not necessary that the computer understands language in any absolute sense. If language does not have this intrinsic connection with the world, then the question of what knowledge is, is open to answers different from those which locate the source of knowledge in the same place as the connection purported to exist between language and the world.

JUSTIFICATION AND KNOWLEDGE

It can be argued that of the things said, some we have the right to be sure of, others we have some rational basis for acceptance, others we find highly doubtful. Those things said for which we have a right to be sure encapsulate our knowledge. A stronger claim is made by Johnson and Hartley (1982) where they draw upon Austin (1946) and claim that knowing is precisely having the right to be sure. This right to be sure need not be found in an intrinsic connection with the world. It can be a more subtle right involving argumentation.

In particular fields of competence, the standards and methods developed provide the criteria for what it is to have the right to be sure. These standards and methods are not the business of the technologist. The role of the technologist is to provide reliable tools for the transfer of knowledge to a computer. These tools include not only software but methods of knowledge elicitation. It is the technologist's responsibility to elicit the knowledge and not the expert's duty to prepackage the knowledge for the computer. The expert's knowledge is not normally communicated in a form, nor understood in terms, that enable a technologist to directly implement it, hence the need for elicitation techniques. Only if the technologist reliably identifies the knowledge can the 'knowledge base' transfer the right to be sure. One of the technological implications of this view is given in Johnson (1985).

When do people have the right to be sure? In an everyday context, people normally give themselves the right to be sure by what they see – they know by seeing. There are, of course, circumstances where their judgement is suspect, and in these contexts there is no right to be sure. We cannot know by seeing at a conjuring show nor rely on our memories after hypnosis. In these and more esoteric contexts the right to be sure can only be established by complex justifications. 'To know' can be achieved by 'arguing'. In mathematics formal proofs are the methods of justification; in other fields the structure of the arguments can be different from formal proof. Arguments in law, for example, have other structures for validation. This may be done by providing supportive, but not deductive, statements that bridge the gap between premise and conclusion. Thus, if we make the statement:

'Harry was born in Bermuda' (1.1)

and conclude that:

'Harry is a British subject' (1.2)

then we must show how statement (1.2) is derived from statement (1.1). The linking statement in this case is:

'A man born in Bermuda will generally be a British subject' (1.3)

However, if Harry turns out to be a horse or both his parents were aliens or he has become a naturalized American etc, then this link can no longer be justified. Further, the supportive statement (1.3) requires further justification (Toulmin, 1958; Johnson, 1983). This mode of reasoning is sometimes referred to as non-monotonic reasoning because the proven conclusions cannot be guaranteed to hold if new assertions are made. Normal logic is monotonic in that deduced results are independent of any new assertions. It is this characteristic of *monotonic* logic that forms the basis of 'rule independence' in expert systems. This will be explored further in Chapter 8.

When is the process of justification terminated? Is there a level of justification that is accepted by any reasonable man? This level has no absolute criteria, and alternative groups have different base premises. An expert justifying a fact to another expert will argue in a different way than to a non-expert. Thus, a mechanism can only represent knowledge provided that the representation contains its own justification at a level accepted by the group of users. Since the level of justification is not fixed for any group and this level can change in time, then the knowledge of the group (if fixed) will degrade if not continually adjusted.

There is a stronger statement that can be made concerning the knowledge of a concept (Johnson, 1983) in that a concept cannot sensibly be decomposed into a set of component propositions. The problem arises in our example of knowing what arguments will declare Harry not British. The idea of being British is independent of many proposed features since they only declare what is typical. 'Knowing' then becomes a condition of being able to justify the existence or non-existence of a concept (such as being British) in any context including those contexts that have never previously occurred. This can often result in all previous specifications being overriden. It is this ability to conceive of a concept without recourse to old methods of argument, and to generate new lines of support, that indicates a true understanding. After all, this is what makes an expert an expert.

Applied epistemics

It is worth considering a different view of machine knowledge: a view that is compatible with the philosophical elements of knowledge and meaning. One way is to consider the machine in the more passive role of an epistemic mechanism. Epistemics is the science of communicating understanding via stored knowledge. The term was originally coined by P. Meredith (1966) to describe the use of text as a medium for transmitting ideas. The purpose of any scientific text is to communicate an understanding of some particular area of study: to somehow project experience through the medium of paper and print. The structure and form of a document is influenced by certain expectations of grammar and prose. It was pointed out by Meredith that the roots of these expectations lie in the domain of literature and the spoken language rather than in the sciences and the technical report. For this reason, those with a classical background will generally avoid using diagrams or tables to aid explanation.

There is the added suggestion that the use of drawings implies an inability to express oneself properly. However, many of the knowledge representation schemes developed depend upon the expressive potential of diagrams; these schemes take advantage of people's abilities to perceive and understand visual patterns.

The document allows thoughts to be expressed in a time-independent fashion. Sentences can be reread, multidimensional relationships can be illustrated compactly and indentations can mark levels of explanation. There exist a range of skills and techniques for scientific writing which differ from those required for literature. It is proposed to broaden the science of epistemics to include any time-independent media such as computers. Communication theory will certainly be an aspect of the study but the important property will be the storage and manipulation of knowledge representation structures for the purpose of providing understanding. Applied epistemics will be the application of technology to the storage, manipulation and communication of knowledge.

The computer media can extend the concept of a document written by a single author to that of the interactive document written by a group of experts. The fact that it is interactive allows the document to be accessed in a manner suited to the purposes or skills of the user. Furthermore, this interactive group document can be under continual review so that the most recent ideas are available to the participating reader.

Thus, we can move from the passive single-author, and fixed paper-based report, to a multiple-author, reactive and continually modifying system of communication. The system becomes more effective when the document is coded in a form that is homomorphic with respect to the knowledge content, and under this condition the authors are participating through the epistemic mechanism of a knowledge base.

The knowledge base

If a particular set of facts is known about the world, then this factual knowledge can (apparently) be increased if various hypotheses (or rules) are known. Facts may be derived using both observed and derived facts through a justified mode of inference. Hypotheses (sometimes referred to as inference rules) typically express general knowledge couched in a formal system of operations. There are two

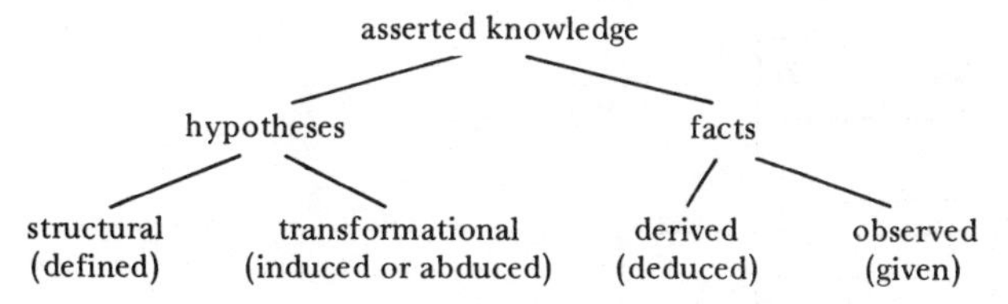

Figure 1.1 *Types of assertional knowledge*

kinds of hypotheses within the context of a chosen mode of reasoning (see Figure 1.1): one that describes change (a transformation) and the other structure (eg taxonomic and categorical structures). Thus, if it is known that a balloon held is lighter than air (an observed fact), a balloon is an object (a structural rule), and that all objects lighter than air will float upwards when released (a transformational rule), then it can be deduced using the acceptable modes of reasoning that the balloon will float upwards if released (a derived fact). Thus 'knowing' also includes hypotheses about the world that can provide an infinite (but closed) source of 'derivable facts'. It is the trace of how these rules are applied within an accepted mode of reasoning (inference) that forms the basis for the explanation.

The knowledge of when or how to apply rules is referred to as 'heuristics'. Heuristics are explicit statements for controlling reasoning. Heuristics cannot necessarily be justified, but reflect an approximation of known constraints.

Modes of inference are a third kind of knowledge that describes how reasoning may be performed under the control of heuristics. In logic, this knowledge consists of tautologies and valid argument frameworks (deduction). However, it can also be any form of acceptable processing such as the set of relational operations. Heuristics, hypotheses and rules of inference normally merge to form a computer program that derives facts (output or intermediate results) from other facts (data).

Computer systems that are organized explicitly and distinctly to represent facts, inferential knowledge and methods of deduction are sometimes referred to as expert systems. Expert systems are normally presumed to capture the essential competence of an expert, and they are practical examples of applied epistemics. Competence consists both of knowledge about the task environment (modelled by structural hypotheses and facts) and the skill to perform the task (modelled by transformational hypotheses and heuristics).

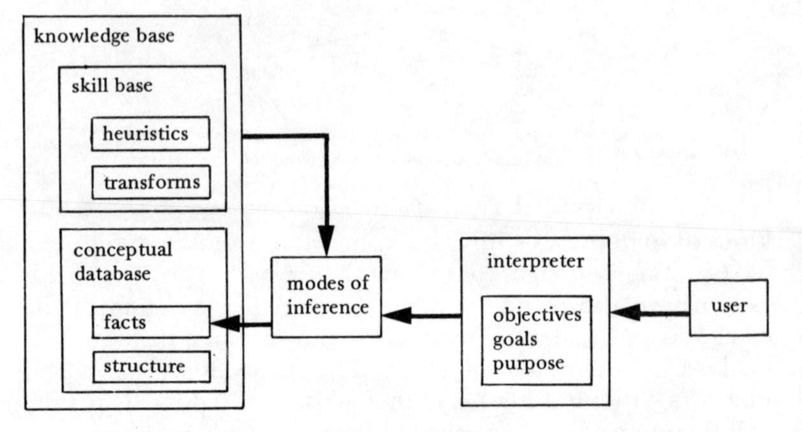

Figure 1.2 *The essential components of an expert system*

An expert system is dependent upon both a knowledge base and an interpreter (see Figure 1.2). The role of the interpreter is to presume a purpose behind any conversation between the system and the user. Thus, a change in the interpreter could convert an expert system into either a management information system or a teaching aid (provided the right kind of knowledge is available in each case). The interpreter defines how a mode of reasoning should be used to achieve the desired ends. In this limited respect, the interpreter is concerned with control by setting the goals of the system. The choice for control is supplied by the knowledge base and the heuristics.

A conceptual database is a restricted form of knowledge base often linked with a fixed set of formal operations (eg relational operations). The conceptual database includes only facts and structural rules. For example, a table has boxes upon its surface (facts), and a pile of boxes is called a tower (structural). The operations associated with the conceptual database are concerned with maintaining specified constraints and deriving predefined classes of facts. The conceptual database describes the static knowledge and indicates clearly how elements relate. It may also describe how things change, but this change is limited to the constraints imposed by the relationships between the facts.

The conceptual database can be extended into a knowledge base by explicitly introducing heuristics and providing a mechanism for a more general form of reasoning (one example is the rule-based inference engine). However, the conceptual database is normally used in conjunction with a set of programs that perform specified tasks. The advantage of a conceptual database is that it presents users (programmers) with a stable interface to a common environment description (ie a description that is always coherent).

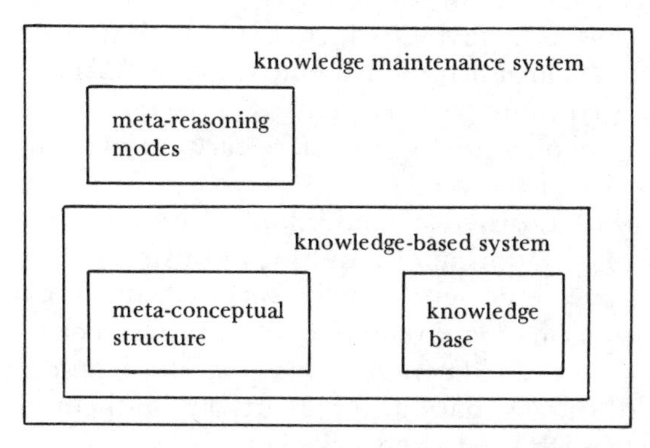

Figure 1.3 *The recursive nature of a knowledge base*

Asserted knowledge descriptions are also facts, and it is possible to have methods of reasoning about them (see Figure 1.3). The meta-conceptual structure defines what is required for a well-formed set of assertions. Meta-reasoning is involved with heuristic strategies for solving problems and extending the mode of reasoning by adding to the control heuristics (rule application). There is no place in this system for determining what constitutes a fact or how particular representation schemes should be interpreted other than through consistency criteria. It presumes a representation scheme but cannot recommend one without reference to the users. The systems record knowledge; they do not contain knowledge.

Knowledge engineering

One of the reasons why a theory of knowledge has been so slow to form is that the nature of knowledge does not fit well into any accepted recognizable theoretical format.
A scientific theory usually attempts to discover a set of finite elementary units and a set of laws of behaviour that allows the construction of models that can recreate the performance of any considered situation. Thus, Newton's Laws together with the concepts of mass and force provided classical

dynamics with a powerful tool for design and explanation. The 'objects' of mass and force cannot be identified as having any existence other than within the framework of the laws of dynamics. They are recognized only by their interaction, and yet have a separate and independent existence. A similar attempt has been made with 'knowledge' in that absolute primitives have been suggested for representation schemes that have laws of behaviour which would be sufficient to explain what is meant for a machine to 'know'. Although such absolute primitives may perhaps exist, they are not sufficient (and may not even be necessary) to explain all the characteristics of knowledge.

A theory of knowledge would thus seem to demand the unsatisfactory condition of requiring both structurally dependent laws and user-defined elementary units. Every possible situation would require a set of typical constraints with its own set of elementary concepts. This would suggest that each of these situations represents a mini-theory of knowledge which has to be worked out afresh according to requirements. An example of a mini-theory is given in Hayes (1979) where he develops a physics for common usage by suggesting primitives that can describe the behaviour of objects and liquids as perceived by our everyday experience.

The questions concerning the new generation of computers are 'What problems will have to be overcome if they are to be truly knowledge processors?' and 'What benefits might accrue once such machines are available?' There seems very little doubt in current thinking that the new generation achievement is not only technically possible and desirable but it is also likely to happen within the next decade. The question arises as to whether these beliefs are justified.

Underlying the discussions in this chapter has been the assumption that the use of future computers will be for real-time decision support systems. Mass processing and other repetitive operations on data (such as payroll) are presumed to take a less important role in the field of computing as real-time interaction becomes available to more people. People will be referring to the computer in much the same way as they might refer to any expert, consultant or advisor. There will be differences in that the attitude towards a computer will be the same as that of the driver towards the car, and as with transport there will be a wide range of possibilities.

A brief sketch of such systems is illustrated in Figure 1.4. This shows that a new breed of computer expert will be required, one who will naturally evolve from the systems analyst, called the knowledge engineer. The task of the knowledge engineer will be equivalent, in many ways, to that of today's systems analyst and programmer. The difference is that the skills will be based upon elicitation techniques and computer technology that are geared to interpreting naturally occurring performance into a formal representation of an appropriate abstraction of that performance (Hartley, 1982), as well as determining the agreed ideal behaviour and practice of the experts, consultants or advisors. One representation that concentrates on the expert's view of the world is that proposed for the conceptual database (see Chapter 4). Such a view excludes modes of reasoning and methods of general problem solving. It is concerned with a set of tasks that needs to refer to the world in a common way. The knowledge engineer will have to acquire a sensitivity to the user's needs and introduce training schedules for the users.

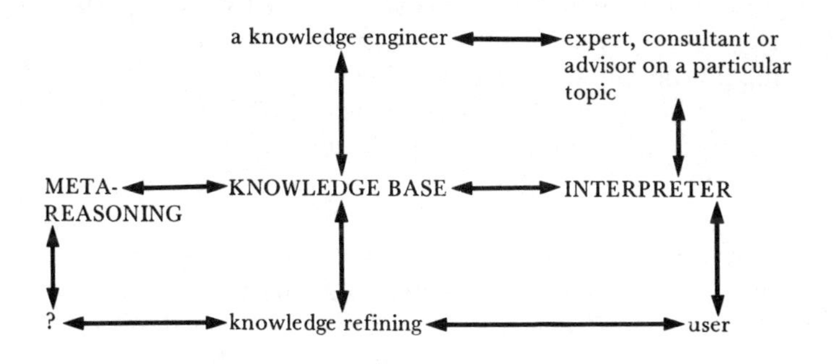

Figure 1.4 *The real-time knowledge-based system*

The lower half of Figure 1.4 illustrates the user being presented with an ordered list of recommendations produced through real-time interaction. The results of this interaction and the success or failure of its use can be employed to 'refine' knowledge. The question mark (?) indicates an important, and as of yet, little understood stage in the process that can generate the required mini-theory of knowledge needed for each task. The problem now is not the generation of specialized programs but the induction of new theories.

It becomes clear that current techniques of knowledge

representation by machine are inadequate for any new generation ideal. Paradoxically, the reason for this inadequacy is due to the concentration on technical problems rather than consideration of the role of the human user within a complete system. This concentration on techniques has been supported by the belief that people are machines. This may well be true, but with only current generation machines upon which to base recommendations of requirements, certain prejudices have naturally followed. The concept that knowledge was similar to data but required a lot of processing turns out to be false. Knowledge representation, it seems, does not depend upon any particular mechanism nor reside in any fixed manner within a machine (Newell, 1980) nor rely upon a unique theory. (Lenat, 1977, attempts to tackle the question mark in Figure 1.4.) The source of support for any mechanism to have knowledge comes from without rather than from within. To construct a system with knowledge requires the ability to elicit the knowledge and to construct an appropriate abstraction at a level of resolution adequate to deal with the set of expected situations perceived by the group. Once achieved, the constructed system requires continual adjustment in order to keep in step with the group. There have been no techniques of any depth that enable this to be done, and there has been little theory to help. The problem now is not technical in the sense of requiring better hardware, although such developments are needed, but a question of obtaining a deeper insight and readjusting to a different point of view. The remaining chapters will examine the components of real-time knowledge-based systems as aspects of an epistemic mechanism and describe a rational approach to their design.

CHAPTER 2

Relational Analysis

q is for princess,
r is for bunny.

Elizabeth Anne Addis (age 2½ years)

Introduction

Systems analysis has been concerned with the problem of specifying data structures and programs for computer systems that match the requirements of certain kinds of industrial and commercial organizations. Although it is not normally so expressed in systems analysis literature, this activity can be considered as a process of defining a linguistic subculture pertinent to the organization under study. One of the most important analytic tools that has evolved from this work has been relational analysis. Relational analysis for industrial database design was originally offered as an alternative to the CODASYL (*Conference on Data Systems Languages*) proposals of the late 1960s and early 1970s (Codd, 1970). The CODASYL proposals were concerned with, among other things, the unification of database design techniques based upon an extension of the methods of COBOL (*Common Business Oriented Language*) programming (CODASYL, 1971). In particular, the CODASYL proposals centred on file design for large-scale random access computer data storage that would be independent of the user's programs. Complete independence was obviously impossible, but it was thought that if the data storage conformed to some unifying principles, then any competent programmer ought to be able to access data from a common pool; and further, that this common pool should be capable of being extended both structurally and in content without disturbing existing programs. In this way an organization could possess a true commercial database to which any reasonable question could be posed and using which any reasonable task performed.

Relational analysis was originally presented as a technique for the design of actual physical files, and the relational database was considered as an alternative to a CODASYL database or other structures. As a design methodology relational analysis had much to commend it and it was claimed that:

1. The relational concepts were based upon a well-formulated and understood calculus. The calculus was theoretically respectable and could be considered complete.
2. The relational design technique remained independent of the physical storage, the computer used, the access strategies employed by the database system and any bias towards answering particular questions.

   (This point was of particular importance since every design technique at that time required an early decision to be made as to the use of the system. The fact that it was independent of physical storage or the computer used indicated that it was not really to be used literally for physical file design.)

3. The relational specification could accommodate change and growth of a database as well as allowing comparison and merging of different databases.
4. The relational structure maintained the logical truths of the universe it represented.

   (This was of interest since file design could not always guarantee to be homomorphic with the data contained within its framework.)

5. A subset of the database could be presented without inconsistencies.
6. The relational concept was considered easy to understand.

Relational analysis, as it was originally presented, could not fully support all these claims. Much of the early work on relational techniques still confused file design with relational analysis, and many of the arguments that were used to support the various stages in normalization theory encouraged this erroneous view by depending upon a physical analogy. This original confusion is still widespread, and the misunderstanding of the purpose behind relational analysis has resulted in its rejection by some computer professionals.

A better use of relational analysis is to provide a method of data description that is unequivocal. Once an analysis has been achieved, the physical (or logical) file design from such an analysis can be a separate process which takes into account the special requirements of the computer architecture, the frequency of access and type of program that will use the database. (This physical stage of file design will be considered more fully in Chapter 4.) However, relational analysis has its own ontogeny, having been born at a time of commercial expansion of computer usage that exposed the inadequacies of the then current design methods. Relational analysis ought to be considered as a response to quite specific problems that, although still present, are no longer of major concern. In the intervening years, technology has moved quickly and has provided its own type of solution to the problems.

Relational analysis as initially expressed by E. F. Codd examined existing files as though they were relations. It was common during the 1960s and early 1970s to store bulk information on magnetic tape. Many of the records within such files would be structured so that they contained repeating groups. A repeating group was a set of fields that could occur one or more times within a single record, thus representing multiple occurrences of objects or entities with the same descriptive features. An example is a PART record and the list of suppliers from whom that part can be purchased. This method of storing information has advantages within the limitations imposed by sequential accessing and a single use of the data. Problems arise when random access methods become available and the multiple undefined user demands of a database are in force.

One of the major problems was the 'many to many' situation where, for example, many suppliers can supply the same part and many parts are purchasable from the same supplier. If the only means of representing data (and consequently the only tool for file design) was the repeating group, then this situation could not be described without introducing a bias. The main record sequence had to be either the part or the supplier. There seemed at the time to be no way of representing (and hence storing) data in a way that reflected this symmetry.

Random access was provided by the introduction of the disk, but the representation schemes were still concerned with repeating groups (sets and set owners). Relational analysis, on the other hand, broke the bonds of repeating groups by the introduction of relational algebra. It no longer became necessary to specify the data in the form of tree structures since relations could interlink sets without bias. Relational algebra defined a set of operations on relations that enabled new relations to be constructed. If the original relations were 'normalized relations', then the relational algebra would be 'well behaved'. Computer files could then be regarded as relations with each record considered as an n-tuple and each domain (or attribute) a record field. All that was needed was for the computer files to be 'normalized' in the same way as the relations and for a set of operations on those files to be provided that corresponded to relational algebra. The result was the 'perfect' database unsullied by representational problems. Every 'normalized' file could be treated uniformly instead of having special access modes for each type of record structure, and a fixed set of operations could be defined in the form of relational algebra to provide all the data manipulation requirements. It was further noted that such 'normalized' files had some useful updating properties. However, it is these updating properties that present a major theoretical difficulty, which will be discussed below.

Relations and tables

A relation may be considered to be a sentence with blanks where the blanks are to be filled with values from a specified domain in order to obtain a true proposition. Thus:

. . . is greater than . . .

and:

. . . is between . . . and . . .

are examples of two and three place relations with all blanks being filled from the domain of real numbers (say). A relation can also be thought of as a predicate in that given a suitable pair, triple or n-tuple of values then the result is 'true' or 'false'. The above sentences may be expressed more briefly in this role by R1(A, B) and R2(X, Y, Z) where A, B, X, Y and Z indicate the domains for a particular place. R1 and R2 represent the sentences.

Given every possible combination of values (ie the cross-product of the domains), there is a subset of n-tuples (pairs, triples etc) for which the predicate will respond 'true'. This subset of n-tuples forms the satisfaction set. The satisfaction set is the 'extension' of the relation, and its predicate is the 'intension'. Thus, the extension is the visible realization of the intension. For the formal definition of a satisfaction set as well as for other forms of representing mathematical relations see Bandler and Kohout (1985). This reference deals with both binary and n-ary relations, as well as with fuzzy relations (see p. 230).

There are two traditional ways of viewing a relation R between two sets X and Y, the *intensional* and the *extensional.* The first is at least as old as Russell (1903), while the second stems from Peirce, is enshrined in Schröder (1895), and was rediscovered by Wiener (1914). The extensional view almost monopolizes the mathematical literature under the influence of Bourbaki although the *intensional* view may better reflect the intuitive process of a designer of a technological artefact (Bandler and Kohout, 1985). For the purposes of mathematical argument, the satisfaction set is used because it is easy to manipulate. For example, two relations are considered equal if they have the same satisfaction set; they cannot be distinguished. Relations are then ordinarily defined as a subset of the cross-product of the domains, and consequently the intension is not used. Hence, if there are two relations R1 [Name, Age] and R2 [Name, Shoe-size] which have the same extension, then according to this extensional definition they are equal (ie the same relation).

There were three additional ideas introduced in order to model a computer file as a 'relation'. The first idea was to visualize a computer file as a 'table' where each row (or line) of the table corresponded to a record within the computer as well as to an n-tuple in a reference relation. Therefore there are three models: the computer file, the table and the relation. The table model and the relation model are intended to be equivalent and differ only in terminology. In practice, the table model acts as a convenient intermediary between the formal and the engineering aspects of database description. The column headings of the table are called attributes and are related to, but are distinct from, the domain of that position in the associated relation. The columns are the attribute values. Domains are considered to be a resource from which attribute values can be drawn; consequently, two attributes may refer to a common domain. It has become a matter of form to give an attribute the same name as the reference domain and maintain distinctions through the name prefixed by its role within the relation. Thus, within a single relation that has two attributes from the same domain (say number) we may write:

R1[Largest-Number, Smallest-Number]

and given the two relations R1 and R2 we may write:

R2 [Middle-Number, Largest-Number, Smallest-Number]

Attributes with the same name and role can be distinguished by prefixing them with the relation name. Then:

R1. Largest-Number can be distinguished from R2. Largest-Number

The second idea was that some attributes are marked as having a special role. These attributes form a minimum grouping such that their values in concatenation can uniquely identify a single n-tuple in a relation (or row in a table). There may be several such groupings, some of which may contain common attributes. These groupings are called KEYs, and one of these groups is chosen as the PRIMARY KEY. All other descriptive attributes are called OWNs. The purpose behind this distinction was to acknowledge the existence of sequential indexing (on the primary key) and inverted indexing (on the non-primary keys) within a mathematical framework.

The third idea was that the table (or relation) could be 'time varying' by allowing new rows (n-tuples) to be inserted, old rows to be removed from a table or values changed in a row. The argument was that the intension of the table (relation) is both the name of the table and the column headings (set of attributes), and is to be interpreted as a predicate that is satisfied by all 'legal' extensions. The extension is the current set of values. [Note that the normal mathematical conception of a relation is that the values are static (the satisfaction set), and many of the formal arguments depend upon this view. Thus, the table is a fundamentally different model to that of the mathematical relation.]

Functional dependency

The process of normalization depends upon the analysis of the functional dependencies between attributes. A functional dependency (FD) specifies a restriction between attribute values that is 'understood' *a priori*. It occurs when the values of an attribute set uniquely determine the values of another attribute set (Maier, 1983). Thus, for the two attribute sets X and Y:

$$X \to Y$$

means that for every $x \in X$ there is a unique value $y \in Y$. Y is said to be 'functionally dependent upon X' or 'X determines Y'.

Functional dependency is distinct from that of a *function* between X and Y. If there exists a functional dependency between two sets, then this suggests that there is a function between them. A function from sets (ie attribute values) X to Y is a relation f between X and Y such that for every $x \in X$ there is one and only one $y \in Y$ such that $(x, y) \in f$. The set X is called the *domain* of f, and the set Y is called the *codomain* of f. This can be written as:

$$f: X \to Y$$

This definition does not allow the existence of elements of Y that are *not* paired with any elements of X unlike the restriction imposed by functional dependency (Fagin, 1977). The elements of Y to which an element of X is paired form a subset of Y called the *image* of f; this is also called the *range* of f. The function f can only be defined if both the domain and the range are themselves defined. In the case of tables, neither the domain nor the range can be defined, and there arise certain problems of how these notions can be translated. However, for tables the concept of a function is ignored, and the *a priori* knowledge of the functional dependencies is called 'semantic'. Exactly what this could mean has never been made precise.

Two important relational operations

In order to describe the significance of normalized relations, two of the operations in relational algebra need to be briefly described. Relational algebra is covered in more detail in Chapter 4.

PROJECTION

Considering a relation as a table, the process of projection is the removal of one or more columns in the table and then the elimination of any repeated rows in the new table. If there was a relation describing the parts obtained from particular suppliers for current jobs thus:

PJS [Part#,	Supplier#,	Job#:]
1	456	32
1	456	45
3	456	32
6	621	45
6	456	32
2	456	45

then various projections are possible. Some examples are:

PS [Part#,	Supplier#:]
1	456
3	456
6	621
6	456
2	456

SJ [Supplier#,	Job#:]
456	32
456	45
621	45

J [Job#:]
32
45

THE NATURAL JOIN

The join is a process through which two different relations can be formed into a single relation. If a relation is considered as a table, then a new table is formed from the two other tables by producing a union of the set of column headings as the set of new columns in the new table. The process of joining is to create the new table by generating its rows through a process of linking tuples from each of the initial two tables on their common attributes. The common attribute values must be equal. Thus, in the two relations concerned with job assignment and employee holidays:

JO [Job#,	Op:]
32	Mary
32	Bill
32	Jack
45	Mary
45	Bill

OH [Op,	Holiday:]
Fred	Apr
Bill	Apr
Jack	May
Mary	May
Bill	June

a new relation can be produced by taking the common heading Op as the link between the two. The rows in the table (3-tuples in the relation) can be viewed as being generated by writing down each row of JO extended by the attribute values of a row in OH that has the same value of the common attribute (or attributes). This is repeated for every occurrence of the common value (or values). This will result in a new relation (JOH) thus:

JOH [Job#,	Op,	Holiday:]
32	Mary	May
32	Bill	Apr
32	Bill	June
32	Jack	May
45	Mary	May
45	Bill	Apr
45	Bill	June

Notice that [Fred, Apr] does not appear in the new relation JOH and consequently the projection of JOH to give OH′ [Op, Holiday] does not reproduce the original OH. However, the join of PS and SJ in the previous example does reproduce the original PJS. This result is significant in that 'non-loss' joins are a desirable property of any set of relations since any sequence of relational operations can then be defined without reference to its information content.

Normalization and the computer file model

The purpose of normalization is to create a set of tables that presents no anomalous results under the equivalent operations of relational algebra and update procedures. Anomalous in the sense that unintended restrictions are imposed by the representation scheme. In particular, 'projection' and 'join' should be a converse and non-loss process. Further, all the 'data' represented by the tables should be obtainable through relational algebra, and in this sense relational algebra is said to be complete. Files constructed with repeating groups can be represented directly by an unnormalized table. These are relations that have at least one complex attribute. A complex attribute refers to a domain of elements that are also tables. There is no restriction on what these element tables may be, but it is always assumed that they are all of the same degree and each contains the same set of attributes in the same position. Further, there is a restriction normally imposed such that any row chosen from any two element tables which share the same key will have the same values for those attributes that are functionally dependent upon that key (or part of that key). The component parts of a complex attribute cannot be manipulated by the restricted set of relational style operations on the unnormalized table; tables that have attributes with 'values' that are sets or n-tuples cannot have these 'values' manipulated by the 'allowed' processes. Consequently, the unnormalized table must be converted into a table containing simple attributes. When this conversion is done the structured elements of the table are exposed to the relational operations.

FIRST STAGE

The first normalization stage is achieved by creating a new table whose set of attribute names is the union of all the non-complex attribute names involved. Thus the table:

PART [P#: Desc, JOB]

in which JOB is a complex attribute where all its elements are drawn from the power set of the table:

JOB [J#, Quantity-Committed: JDesc, Customer#, CName, Customer-Address]

will become:

PART1 [P#, J#: Desc, Customer#, CName, Customer-Address, JDesc, Quantity-Committed]

The power set is all the possible subsets (including the empty set) that can be formed from different selections of elements from a given set. In this case, the set is the set of rows in the table JOB. It is usual that the primary keys of each complex attribute combine with the primary key of the initial relation to form a new primary key. In this way, the new table can be said to be in first normal form. *A table is in first normal form (1NF) if and only if all the domains referenced by the attributes of the table consist of elements that are non-decomposable (ie atomic).* What is considered to be atomic depends upon the task domain.

The tables in 1NF show certain anomalous results during update. The reason for these anomalies can be exposed by drawing a directed graph of the functional dependencies between attributes. Figure 2.1 illustrates the functional dependencies between the attributes of PART1. An attribute or a composite (such as the KEY) on which another attribute is functionally dependent is called a DETERMINANT. Thus P# is the determinant of Desc; P#, J# is a composite and is a determinant of Quantity-Committed; Customer# is

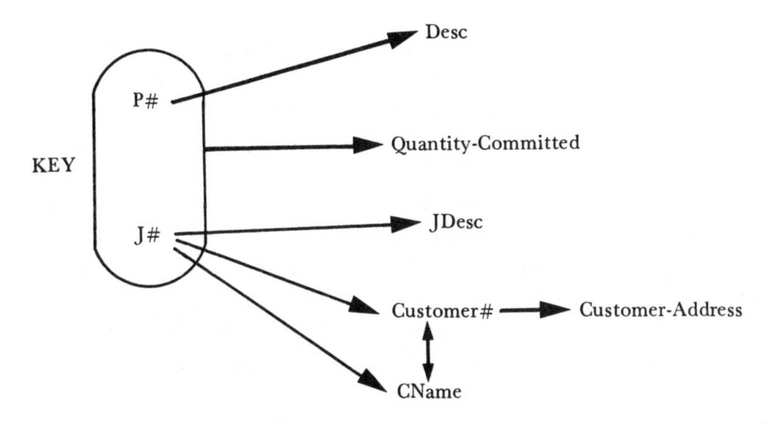

Figure 2.1 *A functional dependency diagram of PART1*

the determinant of CName and CName the determinant of Customer#.

Problems arise because JDesc, Desc, Customer#, CName and Customer-Address depend only on part of the composite key P#, J#. This causes difficulties when any of the values of these partial dependencies needs to be changed because neither P# nor J# independently isolates a single row in a table. This means that the whole table must be searched to ensure that all J# or P# values are consistent. Further, there are no advantages of sequential indexing for the computer file model under these conditions. There is another difficulty in that JOBS cannot be recorded in this table (PART1) unless there is at least one Part (P#) being used on that job. This dependency condition does not reflect the situation to be modelled by the table. However, there is an OWN attribute (Quantity-Committed) that does depend upon the full key P# and J#.

SECOND STAGE

The second stage is to eliminate non-full dependence of attributes on keys by creating new tables. These tables are formed by taking each attribute in the key which has a dependent attribute as the key for a new table. PART2 now becomes the following three tables:

PART2 [P#: Desc]
PARTJOB2 [P#, J#: Quantity-Committed]
JOB2 [J#: JDesc, Customer#, CName, Customer-Address]

A table is in second normal form (2NF) if and only if it is in first normal form (ie all attributes are atomic) and every non-key (own) attribute is fully dependent on the primary key. However, tables in 2NF give similar update anomalies as the tables in 1NF, but for slightly different reasons. JOB2, for example, has information about customers (eg Customer-Address) which is functionally dependent upon the jobs being done for the customer; consequently, if there are customers currently with no jobs, then there are no means of keeping this information. This is the same problem encountered previously. Also every row in the table must be inspected in order to change Customer-Address for a Customer# since a single customer may have many associated jobs. The reason for these restrictions is the transitive dependence of Customer-Address on J# (see Figure 2.1). The transitive functional dependency between J# and Customer-Address bypasses the sequence of functional dependencies between J#, Customer# and Customer-Address.

THIRD STAGE AND BCNF

The third stage is to eliminate transitive dependence of own attributes on keys by creating new tables whose keys are the intermediate attributes. *A table is in third normal form (3NF) if and only if it is in second normal form and every non-key attribute is non-transitively dependent on the primary key.* The example can be re-formed to include a CUST3 table which is extracted from JOB2. However, there is a choice between Customer# and CName as primary key. Their mutual dependencies should be marked so that updates may ensure consistency (no two suppliers with the same name) and also so that the computer-file model can use the index to update CName. There is also a situation where a table may have two composite keys which share one or more attributes (overlapping keys). Since the definition of 3NF is concerned with *non-key* attributes, those attributes that are taking part in a composite key have been ignored and consequently can cause update problems for the same reasons as before. In order to resolve both these criticisms (to mark mutual dependencies and take into account attributes in composite keys), the more restrictive definition of BCNF is given. *A table is in the equivalent of Boyce/Codd Normal Form (BCNF) if and only if every determinant is a candidate key*. It is 'equivalent' because BCNF has only been defined for relations. However, it will be referred to as a BCNF table. The example becomes:

PARTBC [P#: Desc]
PARTJOBBC [P#, J#: Quantity-Committed]
JOBBC [J#: JDesc, Customer#]
CUSTBC [Customer#: CName: Customer-Address]

where the Customer# and CName are candidate keys; Customer# is the primary key. To achieve BCNF may require a further decomposition of composite candidate keys along similar lines to those already needed to achieve 2NF and 3NF.

Normalization and the relational model

BCNF completes the decomposition required to overcome update problems centred around functional dependencies. These problems arose because of the restrictions imposed on the computer file model, by choosing relational algebra as the only usable set of operations, and requiring that access for update should be simple. However, further update problems occur due to the underlying relational model.

THE UNIVERSAL RELATION

The relational model posits the existence of an universal relation. It is required that all relations must be derivable from the universal relation, and, in particular, BCNF 'relations' should be projections. It is through the universal relation that a non-loss set of BCNF tables can be defined since it is known that every combination of values is present. Thus, any attribute value found in one BCNF 'relation' should also be found in any other BCNF 'relation' that shares that attribute. This result should be reflected in the table model so that for example:

The set of JOBBC.Customer# values =
    The set of CUSTBC.Customer# values

This restriction reintroduces the original update problems since every customer must now have at least one associated job and the whole purpose of BCNF (or 2NF or 3NF) vanishes. Normalization makes sense only within the table model: which allows time-varying relations, and where the functional dependencies are constraints on what is meant by a legal extension. A legal extension is one that at least maintains the mappings between attribute values in the universal relation as suggested by a set of functional dependency statements. Hence, if at anytime there was an extension that 'accidentally' gave a functional dependency mapping that was not a member of the predefined set, then this functional dependency is not a constraint and is ignored.

One suggestion to avoid over-constraining a set of tables is to introduce 'blanks' or 'unknown' values within the universal relation so that n-tuples can exist independently in all BCNF relations (and thus tables). However, this suggestion will lose the non-loss guarantee of the universal relation, and unless the distribution of blanks is controlled in some way the purpose of the universal relation vanishes.

FOURTH NORMAL FORM

Indirectly related to the universal relation is another update problem. This is best described through the well-known example (Date, 1981) of a set of courses having been assigned certain texts and tutors. Figure 2.2 illustrates the situation. If there was no desire to manipulate the elements of the lists under TUTORS and TEXTS, then this would be in BCNF. Since the elements of the lists are expected to be involved in the task domain, the table is not even in 1NF and is unnormalized. However, it is desirable to retain the constraint that each tutor of a course should use the same set of texts as all other tutors on that course.

COURSE [Course#:	TUTORS,	TEXTS]
1	(321, 562)	(78, 95, 32)
2	(562, 471, 333)	(21, 33)
3	(471, 333)	(33, 45, 66)

Figure 2.2 *The COURSE table (adapted from Date, 1981)*

The only possible step to normalization is to produce a table whose key is the composite Course#, Tutor#, Text#. The BCNF table is illustrated in Figure 2.3 where all the attributes form the composite key.

COURSEBC [Course#,	Tutor#,	Text#:]
1	321	78
1	321	95
1	321	32
1	562	78
1	562	95
1	562	32
2	562	21
2	562	33
2	471	21
2	471	33
2	333	21
2	333	33
3	471	33
3	471	45
3	471	66
3	333	33
3	333	45
3	333	66

Figure 2.3 *The normalized COURSE table*

The problem that arises is that although the table is in BCNF the rows can only be updated in *sets*. A new tutor for maths requires that the tutor be inserted three times, one for each text on the course (so that he is shown to be using all the texts for the course). It is required that for each value of Course# there is a cross-product of the associated values of Tutor# and Text#. This problem can be avoided by a further stage of normalization where the two interacting multiple sets are separated. The result in this example is illustrated in Figure 2.4.

COURSE-TUTOR4 [Course#,	Tutor#:]
1	321
1	562
2	562
2	471
2	333
3	471
3	333

COURSE-TEXT4 [Course#,	Text#:]
1	78
1	95
1	32
2	21
2	33
3	33
3	45
3	66

Figure 2.4 *Tables in fourth normal form (4NF)*

The dependency that exists between Course#, Tutor# and Text# is called a Multivalued Dependency (MVD). Each value in Course# defines a set of values in both Tutor# and Text#. Thus Course# = 1 defines Tutor# = 321, 562 and Text# = 78, 95, 32 as illustrated in Figure 2.3. The original BCNF table, COURSEBC, can be reconstructed completely (without loss) from these 4NF tables. The initial specification of the problem also stated that a Tutor# could be added to a Tutor# list for any course independently of the associated Text# list. Such a multivalued dependency is indicated by the double-headed arrow with Course# being the MVD determinant. Since in this case it always involves just two other attributes (or attribute composites) this is indicated thus:

$$\text{Course\#} \twoheadrightarrow \text{Tutor\#}|\text{Text\#}$$

A table is in fourth normal form (4NF) if and only if whenever there exists an MVD determinant then all other attributes in that table are also functionally dependent on that determinant in that table. This definition suggests that the underlying table such as COURSEBC contains a multivalued dependency because it reflects (see Figure 2.3) the *join* of the 4NF tables, and, consequently, can only be updated through this operation. A single 4NF table is unaffected by insertions or deletions since an update of COURSE-TUTOR4 (for example) with the row 4, 562 can be done with no consequences. It looks and behaves like any BCNF table. However, if Course# is to continue being a multivalued dependency with Tutor# Text#, then the new value must also be included in COURSE-TEXT4 since a *complete row* must appear in COURSEBC when regenerated through the relational algebra operation JOIN. It may be viewed as though a defined part of the universal relation may not employ 'blanks' or 'unknown' values.

JOIN DEPENDENCY

The non-loss character of the 4NF tables means that they can be joined over their common attribute (Course# in the above example) to reproduce the original table. This principle can be extended to several tables that can be joined in a sequence of any order to form a single table with no loss of information (a loss of information can include extra rows as well as fewer rows than the original). Another well-known example given by Date is the table SPJ describing those suppliers of parts for particular jobs. The important aspect of this example is that there is a form of cyclic consistency known as join dependency (JD) that demands that all three (in this case) pairs of possible projections are needed to reinstate the original table.

The nature of join dependency involving just three attributes (or composites) can be illustrated with a graphical example. This graphical form has not been shown to hold generally. The join of the three projections of SPJ [S#, P#, J#:] naturally defines a loop since:

> If the row ⟨s1, p1⟩ appears in SP [S#, P#:]
> and the row ⟨p1, j1⟩ appears in PJ [P#, J#:]
> and the row ⟨j1, s1⟩ appears in JS [J#, S#:]
> then the row ⟨s1, p1, j1⟩ appears in SPJ [S#, P#, J#:]

and this structure is shown in Figure 2.5(a).

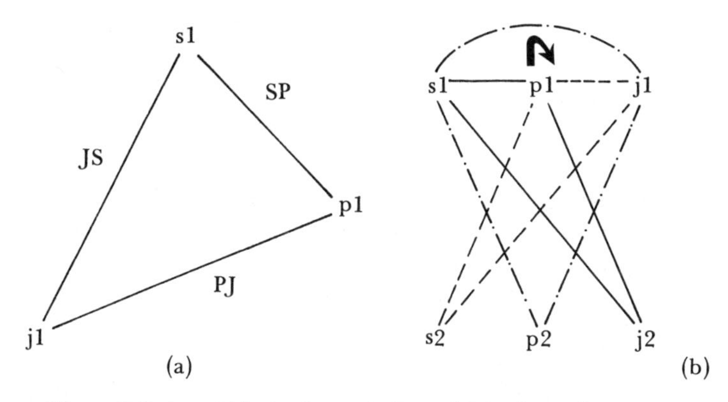

Figure 2.5 *A graphical representation of join dependency*

The 'constraint' of join dependency becomes immediately apparent if it is expressed as:

> If ⟨s1, p1, j2⟩, ⟨s2, p1, j1⟩, ⟨s1, p2, j1⟩ appear in SPJ [S#, P#, J#],
> then ⟨s1, p1, j1⟩ must also appear in SPJ

Figure 2.5(b) demonstrates why this is the case by the drawing of a cyclic path for each triple. ⟨s1, p1, j1⟩ emerges automatically as a direct result of the first three given rows. Figure 2.6 indicates the behaviour under updating of this cyclic dependency. Thus, a row in the table is generated through tracing out every possible loop in the graph. Insert an extra row ⟨s2, p1, j1⟩ and this will also generate ⟨s1, p1, j1⟩ as a new cycle. However, ⟨s1, p1, j1⟩ could have been inserted (link p1 j1) without ⟨s2, p1, j1⟩. Deletion simply requires the breaking of the loop containing the tuple concerned. Thus, the deletion of ⟨s1, p1, j1⟩ can be done by breaking the links s1 p1 or p1 j1 or j1 s1. Each choice will also delete other rows. These will be respectively ⟨s1, p1, j2⟩, ⟨s2, p1, j1⟩ or ⟨s1, p2, j1⟩. The graph simulates the effects of the join.

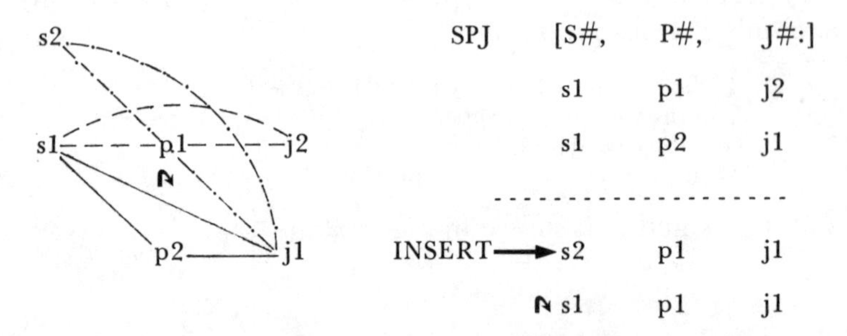

Figure 2.6 *An example of join dependency as a cyclic graph (adapted from Date, 1981)*

A table is in fifth normal form (5NF or PJ/NF) if and only if whenever there exists a join dependency in the table then all attributes in that table are functionally dependent on the candidate keys of the table. That is the join dependencies of three or more projections do not involve cycles (or join dependencies) of the attributes that make up the candidate keys. The JD has been made trivial since it is now reduced to an FD. The 5NF is the last of the normal forms involving projection and join.

Multivalued and join dependencies are unusual in the sense that the definition demands that the complete relation is involved (Maier, 1983). The addition of another attribute which may use the MVD triple, for example, as a key for further description is not dealt with in the current theory. Thus, information about a tutor's favourite text for a particular course would have no place in a relation where multiple update is still required. Suppliers may supply parts at a discount for special jobs (government jobs or jobs done for the trade, etc), and this information is certainly functionally dependent upon the key of SPJ. Yet the existence of such information would appear to destroy the JD since SPJ cannot be simply defined in terms of joins. However, it is not inconceivable to require the effect of the JD on the key of SPJ with an option of additional material about a particular combination once it has been formed. Again, the theory does not seem to allow this simple requirement. Chapter 3 describes a means of containing this problem.

The role of relational analysis

Relational analysis as it has been used for system design straddles three distinct models: the computer file, the table and the relational model. The arguments for normalization are based upon artificial restrictions of the operations that can be applied to the computer file model and the need to have some kind of coherence with the relational view. The strong mathematical foundations of this relational view have been weakened in the table model to such an extent that definitions drawn from the relational model are difficult to justify. The final appeal in the table model is to the 'semantic' aspects of these definitions and the pragmatic notions of simple file organization.

Despite these criticisms, there is an important role that relational analysis plays in that it provides a system design discipline. Without such a discipline, systems may be created on an *ad hoc* basis, at the whim of the designers, without any framework for argumentation. Relational analysis provides a fixed set of operations that forms a central core from which internal design decisions can be made. It ensures that the data objects discovered through the process have well-understood properties. Finally it gives a design theory that may be developed, discussed or, alternatively, rejected in favour of a better one.

Relational analysis (and its three models) focuses on the construction of well-behaved and extendible computer systems. The BCNF relations are suggestive of cognitive objects within a user's mode of operation. The constraints between these suggested objects are not definable within the relational analysis framework because the definitions of normalization view each table that represents these objects in isolation. Relational analysis confines itself to an already expressed environment, and it is only concerned with the partitioning of that expression into these self-contained units. The problems of how the descriptions of the environment may be determined or how the details of the interaction of the entities might be described are not considered. Chapter 3 will examine an extension of relational analysis by proposing a fourth model that takes into account the user's view of the environment.

CHAPTER 3

The Conceptual Model

If things can occur in states of affairs, this possibility must be in them from the beginning.

L. Wittgenstein (1921)
Tractatus Logico-Philosophicus, 2.0121
Routledge and Kegan Paul (1974)

. . . : the meaning of a word is its use in the language.

L. Wittgenstein (1953)
Philosophical Investigations, 43
Basil Blackwell (1974)

Introduction

ABSTRACTION WITHIN AN ORGANIZATION

One obvious feature of industrial and commercial organizations is their tendency to a hierarchical structure of control; it is a feature that becomes more prominent the larger the organization. The inevitability of this structure can be traced to the limitations of people's ability to process and communicate information (Simon, 1969; Forrester, 1975). Larger or more efficient organizations only become possible when these limits are relieved by technology. However, it is not sufficient simply to provide large quantities of processing to improve efficiency, or to increase communication in order to expand an organization. There is a subtle and more important feature of an organization that lies behind this hierarchical structure to be taken into account; it is a feature that relates the form of the knowledge to the activities and purposes of the organization.

It is characteristic of the individuals within an organization to cope by restricting and re-forming information through a process of 'abstraction'. These abstractions represent a mode of communication within a level of an organization, and are formed according to the type of problems requiring to be handled. The distinctions that are made at each level are concerned with the description of the environment in terms that recognize features (attributes) and relationships appropriate to the tasks at hand (the task domain). The environment is thus divided into entities that capture the required granularity of these tasks. There is no absolute set of entities in the sense that some elementary collection of units can be decided upon independently of the context; the entities merely summarize the elements necessary for the work to be done. The relationships between these entities (so created) are sufficient to express the observed restrictions that are functionally coherent with respect to the processes involved. Associated with the entities may be secondary concepts that are formed through constructions made from the primary entities. These secondary concepts produce intermediate structures that are also used for the tasks.

The primary entities, although always potentially recognizable as being complex, are considered atomic within the context of a task domain. Change the task, and what is considered to be an entity may also change. The task domain influences a set of concepts which governs the level of abstraction and selection from the universe of all possible entities. The concepts act as a filter upon the environment; they are governed by the criteria and standards set by the group to achieve the group's purposes. The result is a finite set of entities (not to be confused with the entities referred to in Chen, 1976) that have a common theme, and it is because of this common theme that the atomic entities will have certain interdependencies (mutual constraints). Figure 3.1 shows the relationship between entities, concepts and tasks.

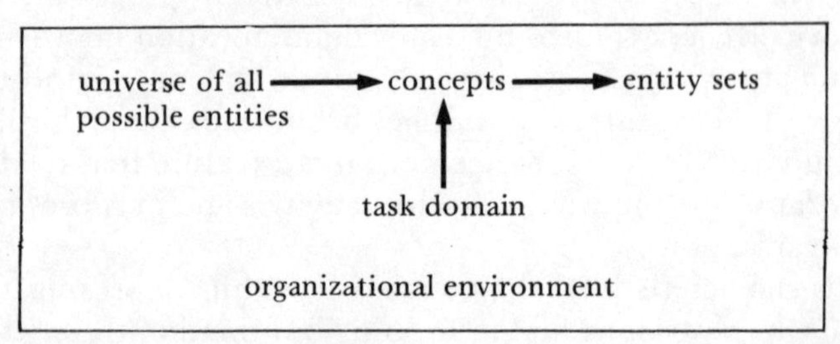

Figure 3.1 *The relationship between entities, concepts and tasks*

Thus, a particular group in an organization will have an empirically based set of perceptions that refers to entities in the group's environment. The entities are abstractions which are dependent upon the intentions of the people, and the nature of these abstractions will to some extent be related to their abilities (Jaques *et al.*, 1978). One of the obvious manifestations of these perceptions is the language and jargon, namely ideas, requests, solutions and information used to communicate within a group. If technology is to participate in the solution of problems, then an analysis technique is required that is capable of describing the perceived environment in terms that are precise enough to be manipulated by computer. Without this analysis having been carried out, the computer has only a passive role.

TRANS-RELATIONAL ANALYSIS

Relational analysis as described in Chapter 2 does not fulfil all the claims made for it in that most of the work on developing it as a technique concentrates on its ability to decompose predefined data into 'normalized' relations. Normalization depends upon the recognition of three distinct models whose purpose it is to justify data independence within the context of both a definable set of operations derived from relational algebra and the efficiency conditions implied by simple computer files. Little effort is given towards solving the problem of eliciting the information from the client (user) or in reconstructing the data to show the necessary transrelational dependencies that reflect the client's behaviour with respect to the data (see footnote 2, page 241, in Date's *An Introduction to Database Systems*, third edition). Consequently, logical truths are not always maintained, and common errors such as the connection trap (see Chapter 4) can occur. Further, although the relational system is presented as tables (not only to overcome the formal restrictions but also to illustrate the user-friendly representation), the method of specifying a program through relational calculus or relational algebra often causes considerable difficulty for non-mathematicians. Despite these confusions the Boyce/Codd Normal Form (BCNF) normalized relations seem somehow to capture the essential entities involved in tasks, but the analysis appears to fail in many other respects.

Work on the ICL Content Addressable File Store (CAFS)

(Maller, 1979a, b) during the early 1970s stimulated extensions to the relational analysis technique that were not published at the time due to the commercial nature of the work (Addis, 1973, 1975a, b, 1976; Babb, 1979; Addis, 1982a; Babb, 1982; Addis, 1983). These extensions were born of a need to express the dynamic behaviour of commercial data independently of any suitable machine that might be used. It was also influenced by the structured descriptions being employed in natural language question answering systems (particularly Winograd, 1972). The natural language work was of particular interest, since it showed what could be meant by 'machine understanding'. This was important because the conceptual gap between information storage and knowledge representation seemed unacceptably wide. However, the kind of complexity normally encountered in artificial intelligence seemed unnecessary in most commercial work. Nevertheless, the techniques developed (and described here) provide a clear-cut approach to providing an analysis of a task domain that gives:

1. an analysis that is sufficiently stable and unique to be used for comparison of systems;
2. a data structure specification that is more precise than had previously been encountered at the time of development;
3. a specification technique such that identical situations will provide similar constructions irrespective of the person carrying out the analysis;
4. an automatic method for generating storage structures (file design) from data structure specifications that can be modified according to local or specialized requirements;
5. a system for accurately specifying all update requirements in order to maintain the logical truths of the environment being modelled;
6. a simple technique for assigning meaning to semiformal queries so that users need never know anything other than attribute names;
7. a method for allowing future changes and modifications to the environment being modelled;
8. a graphical representation scheme that can represent most commercial systems within a single diagram in a way that makes clear its behaviour;
9. a means of testing the elementary nature of perceived entities or concepts;
10. a tool for interview control so that the elicitation of a client's perception of the object types within the task domain is captured efficiently.

The motivation behind the development of these techniques (known as Extended Relational Analysis or ERA) was to evolve a design procedure that fitted the requirements of the analyst, system designer and client (user). The result has been a phenomenological, rather than a formal, approach to design that involves a hybrid representation scheme. It is hybrid because the elements to be captured for the design of a system are of several types.

The three models of relational analysis have been the source of many ideas that helped form ERA. However, it should be emphasized that ERA is based upon different principles and serves a different purpose to that of relational analysis. In essence, the major difference is that relational analysis deals with a well-defined world of attributes and relations; ERA is concerned with the process of discovery of the users' set of perceived entities and constraints which are to be made explicit. Figure 3.2 illustrates the role of each of the four models in the process of design. Each arrow indicates the major direction of influence (cf Figure 3.1). The equivalence sign indicates the step from the users' domain to the computer-compatible representation.

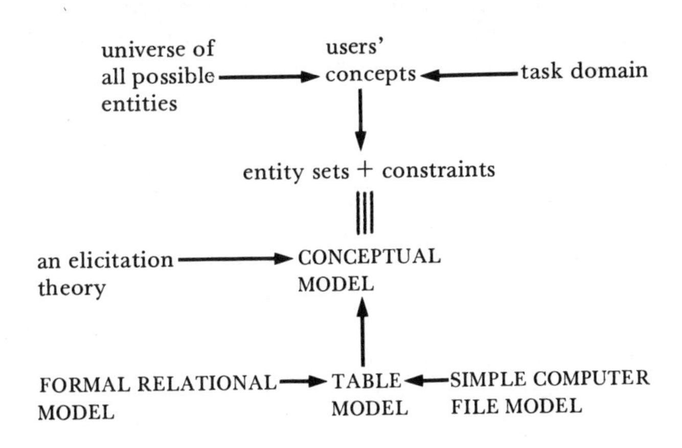

Figure 3.2 *A major influence diagram between models and environment*

Extended relational analysis

INTRODUCTION

ERA is a technique for representing the entity sets perceived by a group of users involved in a particular task domain. In the ERA diagrams, the representation scheme is at a level of detail that makes the interpretation easy and the consequences of the description obvious to the analyst. The description is not sufficiently complete for a detailed implementation specification, but it does capture an accurate account of the users' view of the task domain (Weber, 1976). The result of the analysis is a *conceptual model* (a diagram) that represents the concepts involved in classifying the environment with respect to the jobs to be done. The conceptual model will also represent certain important constraints that exist between classes of entities.* Ultimately, the conceptual model is directly linked to the way in which the users talk about the environment in the context of the performance of their tasks. However, the route from the user's description to the final representation must be taken by an intermediary (systems analyst or knowledge engineer).

The objective of this intermediary is to capture the essential elements of the task domain that are appropriate for computer manipulation without himself needing to be an expert in the domain. This must be done as efficiently as possible. ERA ignores many aspects of skilled human knowledge that might be represented within a computer and concentrates primarily on creating a computer system that can coherently record a complex environment. It provides a method through which users can observe and maintain concrete facts about the world in a form that will be useful.

The table model will be used to represent the perceived entity sets, and in particular the BCNF table will represent examples of a particular concept. The key attributes that uniquely identify a row in a table play a special role since they now also identify an entity in an entity set. The key attributes are frequently artificially created, such as 'Part-number' or 'Supplier-number'. Other attributes are liable to change and thus colour or texture may alter but an identifying number can, and should, remain immutable. Entities are thus identified by the key attribute value and the entity set name. The 'intensional' aspect of a collection of selected BCNF tables can indicate the boundaries of interest for a particular group of users. The boundaries are composed of primitive concepts that are identifiable by simple descriptive

*The conceptual model represents the intensional *not* the extensional description. This is distinct from the normal relational or set theory models.

features. These features may be complex for different circumstances or alternative purposes, but within the task domain under examination they are atomic.

The decision has been made that an entity will have an identity with the option of other descriptive properties. An entity may be abstract, concrete or a relationship. For example, the relationship 'marriage' between the two entities of husband and wife is also an entity with properties such as quality, type and time. An entity takes the linguistic role of a noun, although it may also be used as a modifier (adjective). Thus, colour can be an entity in its own right (eg in the task domain of photography) and can have its own properties such as hue, intensity and saturation. Colour in many situations is also a simple property used to describe concrete objects such as a car.

The conceptual model will consist of a *system* of tables and describes both the entity sets and the constraints that exist between them. The constraints are existence criteria which describe the interdependencies between the entities from different sets. These interdependencies may be due to logical consistency, physical constraints or laid-down protocols. The system defines the limits of what is possible and meaningful to represent about the task domain. To attempt to represent 'facts' beyond these limits is an error. The conceptual model, when implemented within a computer (the *conceptual database*), should respond to manipulation, updating and queries with descriptions that are both possible and interpretable; it is a model of the environment that is set within the framework of abstraction defined by the task domain.

SEMANTIC FUNCTIONAL DEPENDENCY

The table model has evolved from the relational model with an eye to computer implementation. It is because of its strong ties with the relational model that no 'device' could be produced to express explicitly the trans-entity dependencies. It is true that the universal relation suggests constraints, but there is a certain amount of ambiguity in its use. The functional dependencies between the *attributes* of the universal relation are being used to indicate the influence of one *table* upon another. There is an obvious need for something akin to functional dependencies to be expressible for reconstructing the constraints of the environment in terms that explicitly refer to entities. For this purpose it is proposed to say that, when entities in entity sets A and B are associated with each other according to regular and accepted (by the user) constraints of a particular form, then a semantic functional dependency exists between them. Thus, an approach is taken to express the trans-entity structure by taking the liberty of extending the mechanism of mapping to include entity sets. Note that the semantic functional dependency (Sfd) is describing the mapping of *tuples* and not attributes or relations. There may be recognized four basic types of Sfd made out of the primitives 'one-to-one', 'many-to-one', 'into' and 'onto'. These primitives are components of the notation and are:

——	one-to-one	(1:1)
⊢——	many-to-one	(n:1)
≻——	into	('injection')
▷——	onto	('bijection')

The primitives one-to-one and many-to-one may be used by themselves to express relationships between entity sets. Under these conditions the relationships can be given names such as 'sells' or 'owned by'. The names are a linguistic aid to an analyst, and suggest to the designer of a system the common knowledge constraints. For this purpose there is the additional primitive 'many-to-many':

⊢——⊣ many-to-many (n:m)

However, the distinctions which need to be captured for a conceptual model are not expressed by these primitives alone; they must be combined with mapping statements to give Sfds. This inability to make the necessary distinctions, other than through an 'understanding' of the names, can be illustrated by the following examples:

1. PERSONS ⊢—— HOUSE

Here there can be both houses and persons existing independently of each other. The symbol just states that a house entity can have many people associated with it. In this case it is an 'owned-by' relationship.

2. STUDENT ⊢—— COURSE

In this example courses can exist without students, but every student must belong to one and only one course. COURSE here is a complete course of study. The relationship is 'student-of' and must be *given* this extra interpretation.

3. TEACHER ⊢—— COURSE

There may be many teachers for a course but there cannot be a course without at least one teacher. The relationship is 'gives' and again this extra restriction needs to be specified somehow.

0-implies

The 0-implies Sfd is when, at all times, each entity of the entity set A maps into some entity of the entity set B, and there may be some entity of B that has no associated entity in A. In other words, any entity of A uniquely identifies an entity in B, although there may be entities in B that have no corresponding entity in A. This will be referred to as:

A 0-implies B

and will be drawn as a subject/object pair thus:

A >|—— B

The arrowhead (>) marks the subject of the statement, and the bar (|) indicates the potential for zero, one or many entities of A being associated with a particular entity of B. A 0-implies B also includes, as a special case, the notation:

A >—— B

The nearest equivalent way of expressing A 0-implies B in relational terms is through the universal relation, and allowing selected domains (those domains represented in relation A) to have 'blank' or 'unknown' values.

1-implies

The 1-implies Sfd is where, at any time, there can never be an entity of the entity set B without at least one entity of the entity set A, and each entity of A maps onto some entity of B. This is the semantic equivalent to a 'surjection', and is referred to as:

A 1-implies B

and will be drawn as a subject/object pair thus:

A ▷|—— B

or in the special case as:

A ▷——— B

The nearest equivalent way of expressing this in relational terms is by the universal relation with no 'blanks' or 'unknown' values allowed.

Equivalence

If there is never any more than one entity in A associated with any entity in B, as illustrated above, then the bar is not written. This gives the two cases:

semantic equivalent to 'bijection' A ▷——— B
semantic equivalent to 'injection' A >——— B

since the 'bijective' case is symmetrical then this can be expressed as a simple line A——B*. The line is also used to indicate a simple pairing when a and b are specific entities of the entity sets A and B. Figure 3.3(b) shows this use of the line.

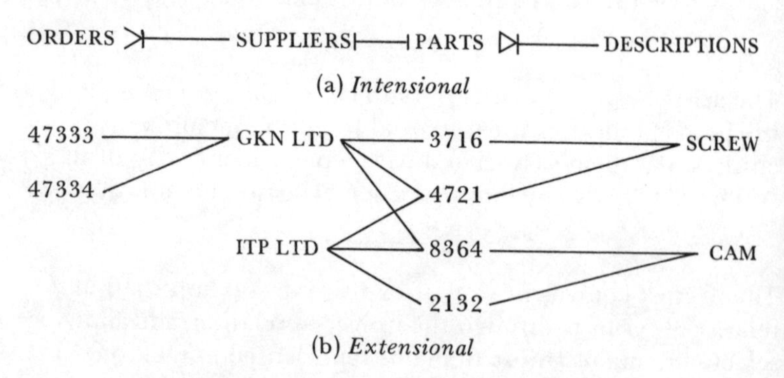

Figure 3.3 *The Sfd 0-implies and 1-implies*

RECONSTRUCTION

Figure 3.3 illustrates two kinds of Sfd between the entity sets ORDERS, SUPPLIERS, PARTS and DESCRIPTIONS. DESCRIPTIONS is not normally considered as an entity set in most task environments. At the time of the initial analysis this role cannot always be eliminated. It is given here as an

---

*Strictly the line is the 1:1 symbol which includes the cases 0:1 and 1:0. If the 1:1 symbol is intended, then an embellished line can be used thus:

A —O— B

entity set because it shows clearly what is meant by 1-implies. The mappings defined are 'intensional' in that they indicate what could be allowed, but not necessarily which values are currently in the table. They do not explain the full notion of the concept since the reason for the choice of values is not given. Since entities are identified by their key attributes then only these are given in Figure 3.3. The current (key) values are part of the 'extensional' description. Figure 3.3 states that every order uniquely identifies a supplier, and there may be several orders going to some suppliers, but not all suppliers necessarily have orders going to them. However, all parts have one, and only one, description, but some descriptions may be used more than once. On the other hand, there is no Sfd between suppliers and parts (only the many-to-many primitive), since some parts might be sold by several suppliers and suppliers may sell several parts. Yet there is a recognizable 'relationship' that exists between them.

Such 'relationships', as between parts and suppliers, may be considered as entities which may have further descriptive features of their own. In this example the entity set PARTSUPPLIER is needed to represent the relationship between PARTS and SUPPLIERS. This 'relationship' has its own attributes, such as the price of a part from a particular supplier, thus illustrating that it represents a concept in its own right. Any entity in the PARTSUPPLIER entity set can uniquely identify an entity in the PARTS entity set as well as in the SUPPLIERS entity set. In this case, it is possible to have entities in either the PARTS or the SUPPLIERS entity set which do not have any corresponding entity in the PARTSUPPLIER entity set. Figure 3.4 shows the Sfd graph between the entity sets.

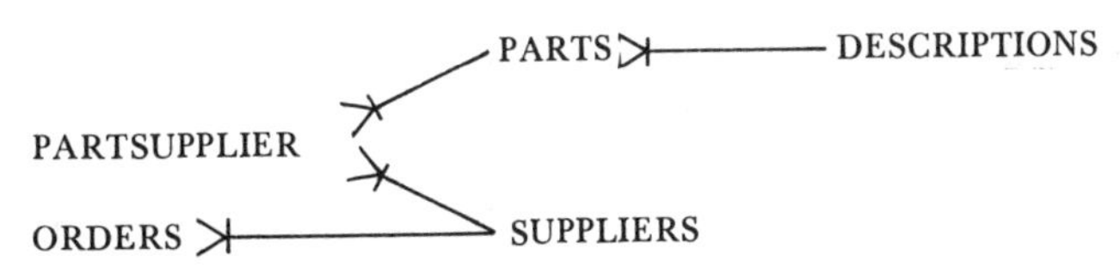

Figure 3.4 *The Sfd graph for PARTSUPPLIER, PARTS and SUPPLIERS*

The practical importance of this reconstruction is threefold:

1. It describes precisely which tables have to be checked during any update (see Chapter 5). Thus, if a new PARTSUPPLIER entity is

to be recorded, then both PARTS and SUPPLIERS tables must be scanned for the existence of the appropriate Part No and Supplier No to be used. Alternatively, if a SUPPLIERS entity reference is to be deleted from the table SUPPLIERS, then any PARTSUPPLIER or ORDERS entity reference dependent upon it must also be deleted.

2. It forms the basis of any file design decisions since any physical distribution of the data ought to maintain the dependencies given (see Chapter 4).
3. It provides a primitive model for an interactive language understanding system (see Chapter 5).

Representations of a conceptual model

Sfd CONSTRAINTS

A set of environmental constraints for a commercial database can now be represented by a semantic functional dependency graph with table names as nodes (Sfd graph) and a list of BCNF tables with their attribute specifications. A computer implementation of a conceptual model will be called a conceptual database. Provided that the implementation remains homomorphic to the specification then the dynamic behaviour of the system can be understood and followed by the user and designer. This has the great advantage of avoiding unnecessary implementation details, which often obscure the processes involved when describing an analysis to the client. An example of such a system is the purchasing system illustrated in Figure 3.5. In this graph each node represents a table. Thus, the relations involved are PARTS [PartNo: Number-in-store], SUPPLIERS [SupNo: Address] and CATALOGUES [PartNo, SupNo: Price], etc. Each table consists of rows, each declaring a fact about the environment. In this example DESCRIPTIONS has been dropped because its linguistic role is a modifier of PARTS in the task domain. Hence DESCRIPTIONS is just a property (attribute) and *not* an entity set (as was initially presumed).

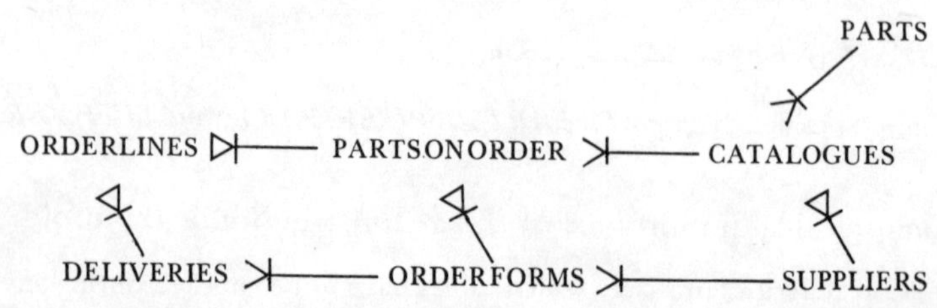

Figure 3.5 *Sfd graph for a purchasing database (a conceptual model)*

The Sfds give the constraints between the entity sets that must be maintained by update procedures in the conceptual database. Thus, if a new catalogue item is to be inserted into the table (or at least the computer equivalent of the table) CATALOGUES, then a part in PARTS and a supplier in SUPPLIERS must also exist to correspond with that item. On the other hand, if an item is to be deleted from the CATALOGUES, then all elements in PARTSONORDER (which contains a list of all things on order) and ORDER-LINES (which is an item on an order form) that imply that particular CATALOGUES item must also be deleted. If the entries in these tables are not deleted, then there will be references to parts that are no longer catalogue items. Further, if the catalogue item is the only item supplied by the supplier, then the 1-implies insists that the supplier must also be deleted. This is strange behaviour, but it is the result of statements made by users (this was an actual case, Addis, 1973). The defects of the system are immediately exposed by the Sfd graph on 'running' various situations through it and checking with the user to see if the behaviour is correct. In this case the problem can be resolved by preventing deletions in CATALOGUES (see p. 61) and by including a 'start-and-finish-date' pair of attributes in this entity set.

In short, the insertion of a new piece of information requires the checking of related material in the implied tables, and the deletion of information requires the elimination of all dependent lines. The Sfd graph gives these update constraints. It is through the knowledge of these constraints, the many-to-one or one-to-one characteristics of the 0- and 1- implies as well as information about usage, that computer file design decisions are made. The Sfd graph has a further use by acting as a reference for query interpretation. It is this query use of the conceptual model that bridges the gap between the user's view of the problem and the activities to be performed by the conceptual database.

CARDINALITY CONSTRAINTS

One of the difficulties encountered during the early development of ERA was the necessity to express other kinds of constraints. One of the most frequent problems was the need to express 'cardinality'. 'Cardinality' is a restriction on the subject set that is described in terms of the individual elements of the object set. If there is presumed to exist an Sfd from A to B such that at all times each value of A maps into some value of B, then we may express an inverse constraint on A such that for each value of B there exists a subset of A whose cardinality (number of elements in the subset) is restricted. This restriction may be written (in words), but it is useful to have a means of expressing the restriction symbolically. The following symbolic representations are intended to be a shorthand way of describing the restrictions, and should not be taken to have the full force of a formal equation (see Chapter 8 for a summary of predicate calculus).

0- and 1- implies are cases of cardinality and both can be expressed by the equation:

$$\mathrm{CARD}(A, f, B) \Rightarrow (\forall\, b \in B)\{(A \supseteq \overleftarrow{f}(b))(n \leq |\overleftarrow{f}(b)|)\}$$

where n is 0 or 1 are special cases, and CARD indicates that it is the cardinality constraints which are under consideration. When $n = 0$ the RHS is trivial. $|\overleftarrow{f}(b)|$ is used to represent the number of elements in the subset of A that are associated with a particular b [the cardinality of the 'reverse' element-set (semantic) function f]. Although the term reverse element-set $\overleftarrow{f}(b)$ was used by Bandler in his papers on relations, originally, a better name for it was coined by Bandler and Kohout in 1976, namely 'foreset'. The following definitions are taken from Bandler and Kohout (1980):

1. The *afterset* of $a \in A$ is $aR = \{b \in B / aRb\}$
2. The *foreset* of $b \in B$ is $Rb = \{a \in A / aRb\}$

where the slash (/) denotes 'such that'. The foreset and afterset of relations are important for representing knowledge (Bandler and Kohout, 1980) in information retrieval (Kohout *et al.*, 1983) and in the design of relational machine architectures (Kohout and Bandler, 1985). The symbols denoting the foreset cardinality state that the constraint of n-implies is such that there exists a single semantic function that will apply to all elements of B and that the reverse of this function for each element of B will form a subset of A of cardinality greater than or equal to n (and may be empty, namely $n = 0$).

More complex examples can be given as in a STUDENT/CLASS conceptual model. In this model a CLASS entity cannot exist unless it has at least four STUDENT entities and no more than 40 STUDENT entities associated with it. Figure 3.6 illustrates how this may be written.

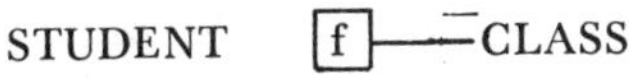

STUDENT [f]——CLASS

$$\mathrm{CARD(STUDENT, f, CLASS)} \Rightarrow$$
$$(\forall y \in \mathrm{CLASS}) \{(\mathrm{STUDENT} \supseteq \overleftarrow{f}(y)) \wedge (4 < |\overleftarrow{f}(y)| < 40)\}$$

Figure 3.6 *STUDENT/CLASS cardinality constraint*

These constraints may be further enhanced by entity-specific limitations. For example, although the above constraint is true for most CLASS entities, some may require that there be only an even number of STUDENT entities as for dancing or fencing. Alternatively, some class entities may require different ranges of cardinality or some unique predicate (C) that declares entity existence. Under these circumstances special cardinality attributes have to be added to the tables so that the cardinality relation will contain variables:

$$\mathrm{CARD(STUDENT, f, CLASS)} \Rightarrow$$
$$(\forall y \in \mathrm{CLASS}) \{(\mathrm{STUDENT} \supseteq \overleftarrow{f}(y)) \wedge (A < |\overleftarrow{f}(y)| < B) \wedge C\}$$

where A, B and C are attributes of CLASS.

UPDATE RESTRICTIONS ON TABLES

In some cases an update constraint needs to be specified which restricts the ability to delete or add rows to a table. This restriction can be marked by prefixing a table name (R) with symbols where:

−R indicates a table R which can only be deleted from. This represents an irreplaceable resource that is drawn upon such as allowed order numbers or the paintings of Constable.

+R is a table R which can only be added to and represents an accumulation of entities. Such a restriction may be required in order to prevent inconsistencies in that the deletion of a row can generate unwanted effects throughout a model, and it may be more appropriate to assign a termination attribute which makes those objects no longer needed.

#R shows that a table R can neither be deleted from, nor added to, although there may be no restriction on changing any values of the own attributes.

Tables which are unmarked are assumed to be completely updatable and to have no restrictions other than those imposed by their position in the conceptual model.

The effects of multivalued and join dependencies on ERA

REVISION OF THE ENTITY MODEL

It was presumed in this chapter that the elementary unit of discourse for a particular group of users involved in a specific endeavour was the entity, and that this entity could best be modelled by a row in a BCNF table. The intension of the BCNF table may be interpreted as the concept that is given form by the extension of the table. However, the individuality of an entity of this kind is in question when multivalued or join dependencies occur.

JOIN

As was shown in Chapter 2, a specialization of the join dependency is the multivalued dependency. The join dependency implies a constraint that should be made explicit, and to this end a new optional symbol of a JOIN can be added to the graphical representation of the conceptual model. However, the JOIN is an operation on tables (or relations), and it is independent of the Sfd description. It describes a further detail about the active constraints. The ERA model now includes tables in other normal forms provided they are at least in BCNF. The important aspects to be retained are the uniqueness of an entity, its potential to have further properties and its position within a framework of existence constraints.

The need for further normalization, as such, is not required but there is a need to recognize multivalued and join dependencies acting upon the *keys* of tables. To take a simple example, consider the problem of generating a table of football matches to be played by a set of teams. Each team must play every other team, and, consequently, the cross product is produced (teams that play themselves are either removed or it is interpreted as a trivial case; see also p. 65). This is a form of multivalued dependency in that if a new team is added to the list, then the table of matches must be increased by the addition of every pairing of the new team with the existing teams. However, it is also required that the time and place for each match be recorded. Since this relationship between teams is an entity set and the key is well defined (a pair of team identity numbers) then this additional information can be added. Entity sets thus remain represented as BCNF tables by dealing with the keys separately from the own attributes. Both multivalued dependency and join dependency are treated in a similar manner when the 4NF and 5NF procedures are used to ensure that the component entities that govern update are correctly formed.

Figure 3.7 shows the JOIN symbol as a double tie between two 1-implies Sfds. The Sfd statement still holds, but the join provides additional information. This is to be interpreted such that any row in the COURSE-TUTOR-TEXT table has a key constructed from the natural join over the common attribute (Course#) of the two tables COURSE-TUTOR and COURSE-TEXT. If there is no common attribute or designated pairing of attributes, then the join reduces to a cross-product (marked by a 0). Each row in the COURSE-TUTOR-TEXT table represents an entity, and, consequently (unlike the equivalent relation), it may have properties of its own (own attributes). The definition of 4NF does not allow this latter possibility (as discussed in Chapter 2).

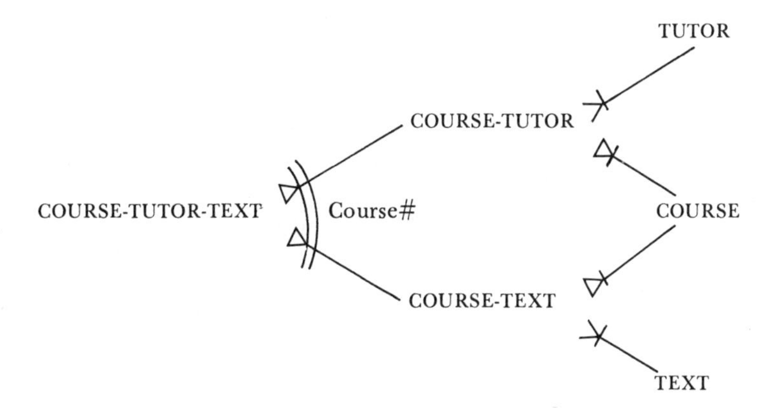

Figure 3.7 *The JOIN operation used to illustrate multivalued dependencies*

The Sfd graph clearly illustrates certain consequences of the multivalued dependencies. The 1-implies from both COURSE-TUTOR and COURSE-TEXT to COURSE restricts the two tables to sharing the same set of courses so that there will be no loss of key information when the two tables join to form the keys for COURSE-TUTOR-TEXT. The 1-implies symbol used with the join symbol indicates that the 1-implies property still holds and is redundant in this case.

Figure 3.8 shows a multiway join where there is a join dependency constraint. The cyclic condition illustrated in Chapter 2 may be inferred from the 1-implies since each row in each pair of tables from the three SUPPART, PARTJOB and JOBSUP must share a common set of attribute values that makes up the key. As with all models of an entity each table may be further embellished with own attributes.

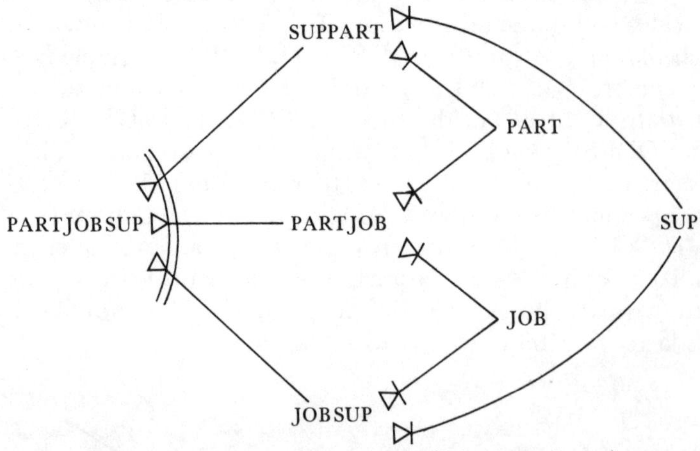

Figure 3.8 *The multiway JOIN (join dependency)*

The double tie also defines how updates will occur in the conceptual database such that the key of the subject table (COURSE-TUTOR-TEXT or PARTJOBSUP) can only be created from the object tables by the process of join. Therefore, updates of the subject table can only be done indirectly via the object tables. To include a new PARTJOBSUP, for example, will require insertions in SUPPART, PARTJOB and JOBSUP first. This will then generate the *set* of new keys for PARTJOBSUP. Once the keys have been established, the own attribute values of the new subject lines may be added.

INHERITANCE
Inheritance of properties is where entities can expand their own attributes by adopting attributes from entities of a different type. This adoption process can only occur within certain well-defined conditions. Thus, the owned animal 'Fido' can inherit the characteristics of the animal 'dog' where:

ANIMAL [Animal: Name, AType]
and ANIMALTYPE [AType: Number-of-Legs, Taxonomic-Class]

Inheritance occurs between the subject and object of 0- and 1-implies such that the subject inherits the attribute/value pairs of the associated object. The Sfd graph defines the route through which inheritance takes place as a process of joins. Thus the animal Fido will inherit the properties of animals of his type through the process of creating a new table from the join of ANIMAL and ANIMALTYPE. This new table is no longer in BCNF (and can be marked on a diagram with the prefix *). Similarly, characteristics of the tutors can be associated with the courses each teaches by the join of COURSE-TUTOR with TUTOR.

One of the purposes of the Sfd graph is to define which joins can take place without generating errors. However, joins that occur outside these defined routes may provide information about potential objects (see Chapter 4).

ENTITY RESTRICTIONS
Join describes a class of restrictions on the existence of an entity that depends in a specific way upon other entities. There are other restrictions of this kind that cannot be expressed as a join,and like join may provide additional information about the task domain. For example, there is the case of an entity that describes a part-subpart link (COMPONENT) where the COMPONENT entity set 'double' 0-implies the PARTS entity set. A double 0-implies is where every entity in A uniquely identifies at least one and at most two entities in B at all times. (Other combinations of 0- and 1- implies are possible.) This may be written as:

or more simply as:

A ⟩⟩⟶ B

Where there is a COMPONENT [Part#, SubPart#: Quantity] relation, for example, then this would be written as:

COMPONENT ⋊⋊——PART

However, there may be the restriction that no COMPONENT entity can have a Part# and a SubPart# of the same value. This is written as a single tie with bar thus:

COMPONENT PART

or more simply as:

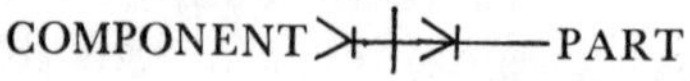

COMPONENT ⋊┼⋊——PART

Restrictions of any kind that are associated with the manner in which attributes come together as entity descriptions may be written next to the single tie (or double tie). The Sfd must have a single object entity set, but may have more than one subject set. Restrictions of this type must *not* be confused with partitions (see p. 67) where there are multiple subject and single object sets.

PARTITIONS

The principle of inheritance can be used to express finer distinctions and represent entities that have much in common, except in certain respects. An example is the entity set represented by the table PERSONNEL that has attributes shared by the table TUTORS and STUDENTS. However, each student may have both a 'home' and a 'digs' address, whereas each tutor will have just a home address. This distinction can be retained by 'factorizing out' of PERSONNEL the tutors and students so that all common attributes are inherited. However, it is also the case that tutors and students may be non-overlapping groups of individuals (say). This PARTITION of a table is illustrated in Figure 3.9 by the single tie with bar within the single symbol, and can be easily distinguished from entity restrictions. The distinguishing attribute is written next to the tie. Note that the form of the compound symbol is like a large injective Sfd but encompassing two subject tables since only a subset of the PERSONNEL can be accounted for in either TUTORS or STUDENTS. The direction of update (insertion) is from the object table (PERSONNEL) to the component subject tables since the distinguishing feature has to be established.

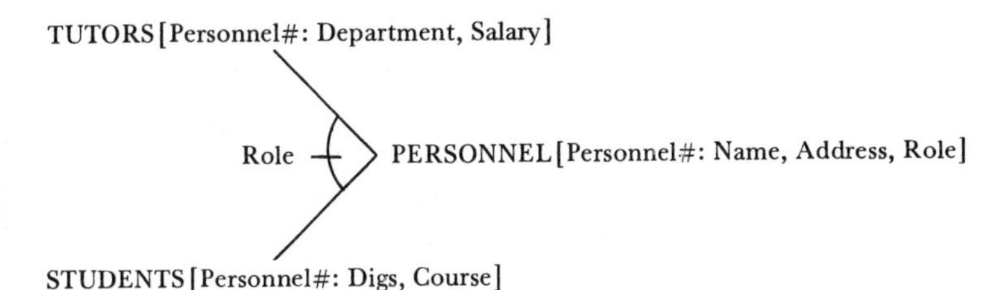

Figure 3.9 *The PARTITION of PERSONNEL on role*

A more complex example of the use of the PARTITION is found in an accommodation problem. Consider the situation of distributing suitable rooms to single or married students with no children. The information required for the task is who is married to whom and the contents of the two BCNF tables:

STUDENT [P#: Name, Sex, Status]
ACCOM [AC#: Type, Address]

It is the rule of the establishment that married couples must have double rooms and single people have single rooms. Figure 3.10 illustrates these conditions as an Sfd graph with partitions. Note that not all the intermediate tables (which have form) need to be expressed, and that, for clarity, the bijective line has been used to link the separate Sfds.

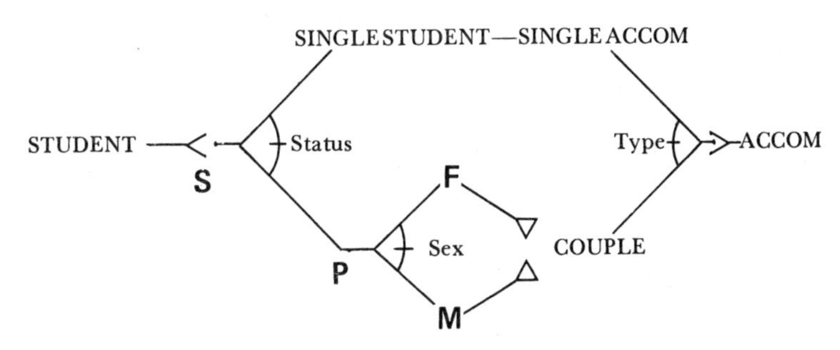

Figure 3.10 *The use of PARTITION in accommodation assignment*

The subset of students as marked by the 'injective' 0-implies (at S) is PARTITIONED according to status. For those of Status = 'Married' there is a further PARTITION according to Sex. On a simple one-to-one basis the two 'bijection'

1-implies form the object table COUPLE that is one-to-one with the PARTITION (Type = 'Double') of the subset of ACCOM. In this case the subject/object distinction is marked to indicate the flow of analysis. For those of Status = 'Single' the object table SINGLE is formed with the PARTITION (Type = 'Single') of the subset of ACCOM. The subsets of ACCOM and STUDENT are chosen so that there can be recorded unassigned entities.

Note that the PARTITION as a meta-symbol is analogous to the 'bijection' in that there is a one-to-one association with the object set partitioned and the union of the resulting partitions. For every parent object there must be one and only one child subject. This can be clearly stated as:

PARTITION ((A,B),C) is such that:

$$A \cup B = C$$

$$A \cap B = \phi \text{ (the empty set)}$$

where A and B are the subject entity sets and C is the object entity set.

ERA as an elicitation technique

ELICITATION AND ACQUISITION

There is a clear distinction within ERA between elicitation and acquisition. Elicitation is the process of determining the Sfd graph via interviews with the client (users). Acquisition is the process of collecting the detailed information (facts) that fits into the framework defined by the Sfd graph. Elicitation determines the intension, and acquisition compiles the extension of the conceptual model. The Sfd graph (and its prototype forms) plays an important role in the interaction between the client and the analyst. The process of defining tables and sketching a graph with the client gives a clear purpose to the discussion. This purpose, from the client's point of view, is to create table definitions and an accompanying diagram that restricts the analyst to making only true or correct statements about the task domain. The purpose, from the analyst's point of view, is to generate statements from the Sfd graph (or some primitive conceptual model) to check its correctness. These statements should be implied through an interpretation of the Sfd graph (or its prototypes). The overall purpose is to define a system of tables (ie tables and constraints) that can record facts about the task domain.

AN ELICITATION THEORY

The system of tables should limit its potential for representation to only those facts that can be true (ie the potential extensions). In this sense the construction of an Sfd graph is analogous to the creation of a scientific theory (see Figure 3.11). The methods of scientific theory formation (The Methodological Falsification Programme) of I. Lakatos (1978) suggest that: a series of theories should be appraised rather than each theory in isolation. An advance is made if a new theory can encompass more known observations and account for a greater refinement in observational detail than its predecessor as well as suggesting new and eventually corroboratable facts.

the Lakatos Programme ——► conceptual model

Figure 3.11 *The influence on elicitation technique*

The Lakatos Programme proposes that an initial theory H1 is supported by a set of corroborating facts CF1. The theory H1 predicts (or can account for) a further set of excess facts EF1 which are either uncorroborated (ie have not been shown to be the case) or are anomalous (ie have been shown to be definitely not the case). This situation is shown diagrammatically in Figure 3.12. The unproblematic background

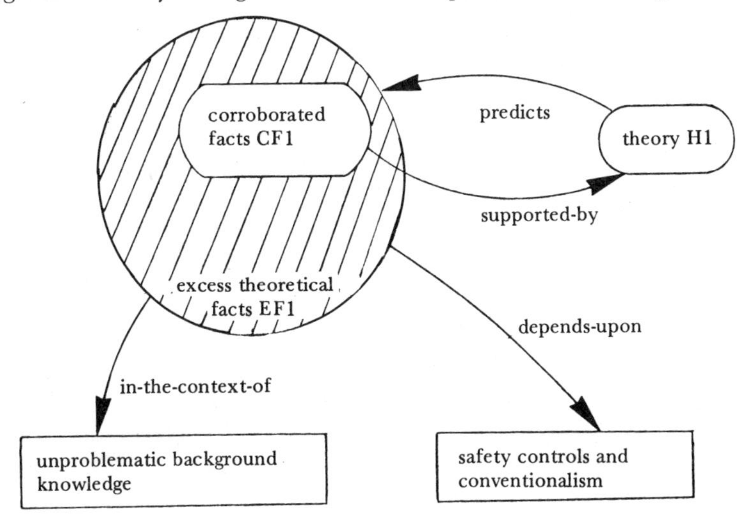

Figure 3.12 *The initial stage in the development of a theory*

knowledge consists of the undisputed facts for which a theory is not required. The safety controls and conventionalism are the agreed methods both of recognizing an observation to be a 'true' fact and of reasoning about these facts once established.

This situation will remain stable until a new and better theory H2 is put forward to account for the anomalous facts in EF1. There are two kinds of theory that may be proposed. The first is a progressive theory PH2 that is supported by H1 and where PH2 implies H1. This progressive theory will be supported by corroborated facts CF2 that include CF1, will account for some anomalous or uncorroborated facts EF1 and will extend the observations beyond the original set of considered facts. PH2 will also suggest a further set of excess theoretical facts EF2. Figure 3.13 illustrates this progressive situation. H1 is only 'falsified' in the light of PH2 since facts can only be interpreted within some theoretical framework. Facts may only be explained or considered relevant with respect to a theory. It is the theory that assigns meaning to an observation, and a change of theory will give observations a new interpretation. A progressive scientific programme is characterized by a series of different revolutionary theories where each stage represents further insights into the environment under examination.

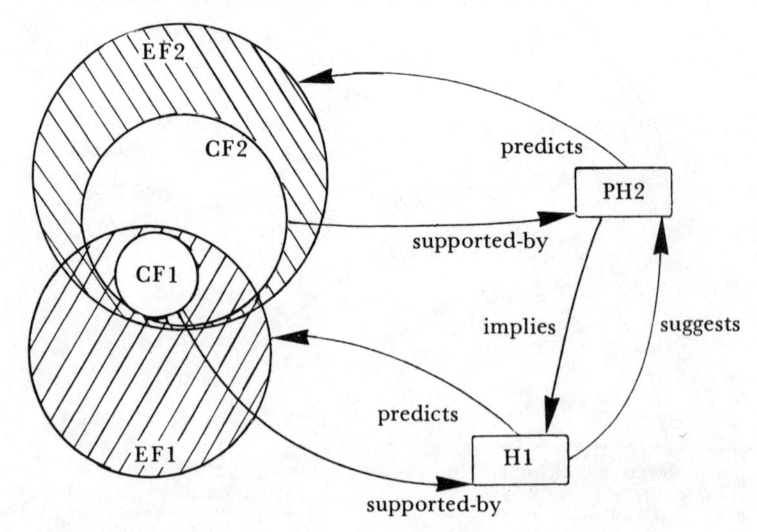

Figure 3.13 *A progressive scientific programme*

The second kind of theory is a non-progressive theory NH2. In this case NH2 is an extension of H1 through the addition of auxiliary hypotheses. Auxiliary hypotheses are extra rules, restrictions, exceptions and modifications to a core theory (H1 say) that refine the theory to take into account only the corroborated (or potentially corroboratable) facts. The result is to reduce the excess theoretical facts to a subset of EF1. This is illustrated in Figure 3.14.

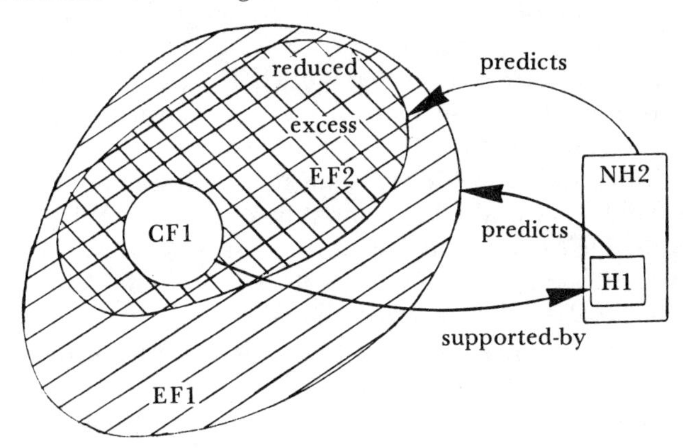

Figure 3.14 *A non-progressive scientific programme*

A scientific theory is not rejected on the results of a single experiment simply because the theory cannot account for an isolated fact. A theory is only overthrown in the light of a 'better' one. Theory H2 is better than H1 if H2 accounts for more observations with greater commitment (eg precision) than H1 in a manner that is progressive. A theory has progressed the series of theories if it can suggest new potential (unobserved) facts (eg considers new situations or old situations with an alternative emphasis).

In the process of elicitation, the ulimate corroborated fact is the case or example supplied by the client. However, it is not practical or even possible for every case to be supplied as evidence to support a theory. The corroborated facts are characterized by the statements made by the client, and these statements can be considered as a theory in a different form to the conceptual model. Such a theory, made of a set of statements, will also suffer from excess theoretical anomalous facts, and one of the objectives, during the creation of an Sfd graph, is to improve the precision of these statements. The Sfd graph is an abstraction and summary of all the statements (made by the client about the task domain) that specify a conceptual model.

A PROGRESSION OF CONCEPTUAL MODELS

The development of a conceptual model will follow both the progressive and non-progressive paradigms. The following paragraphs will discuss and describe the progressive and non-progressive elements of an elicitation protocol as employed to produce the final conceptual model. The use of these elements is geared to the production of new insights and auxiliary hypotheses in an orderly and controlled manner.

The initial model will be very inexact in that many situations will be allowed which would be rejected by the client as 'impossible'. Each model, as it progresses, will call upon more precise symbols that have tighter definitions. Thus, the initial analysis will be a simple listing of the attributes involved without any reference to their interrelationships. Some of these attributes will be recognized to occur in groups that tend to be repeated for each major group of observations. To make this clear, consider the student enrolment problem (originally given by the Open University M352 Block 2). In this example the students are enrolling onto courses. Each student may attend several courses, and each course is taught by a teacher. A student is also assigned to a counsellor who is a member of staff but who does not teach. The students' addresses are required for the records.

The initial 'model' is a set of tables that provides a means of recording the facts. In this example, there is just one table that contains a repeating group CT where:

> CT [Course#, Course-Title, Teacher, Teacher-Staff#, Teacher-Address]

and the table is:

> STUDENT [Student#, Student-Name, Student-Address, CT, Counsellor, Counsellor-Staff#]

At this stage there is no attempt at determining the entity sets or the identifying attributes (keys). The model can define all the required facts (the corroborated facts) but can also express, in extension, many 'facts' that would not be considered possible by the client. Students can have many numbers or counsellors, for example. There are also certain restrictions that suggest that counsellors and students can only exist provided there is at least one of each to be recorded together. The process of analysis which establishes the entity sets is normalization (see Chapter 2). This process is essentially progressive and can be summarized as follows:

1. Remove all the repeating groups and establish identifying attributes
In the example, two first normal form (1NF) tables emerge thus:

STUDENT-1 [Student#: Student-Name, Student-Address, Counsellor-Name, Counsellor-Staff#]

ENROLMENT-1 [Student#, Course#: Course-Title, Teacher-Name, Teacher-Staff#, Teacher-Address]

2. Investigate the functional dependencies on identifiers
The functional dependencies are considered only for the identifiers at this stage, and all other functional dependencies are ignored for the moment. Figure 3.15 illustrates these functional dependencies.

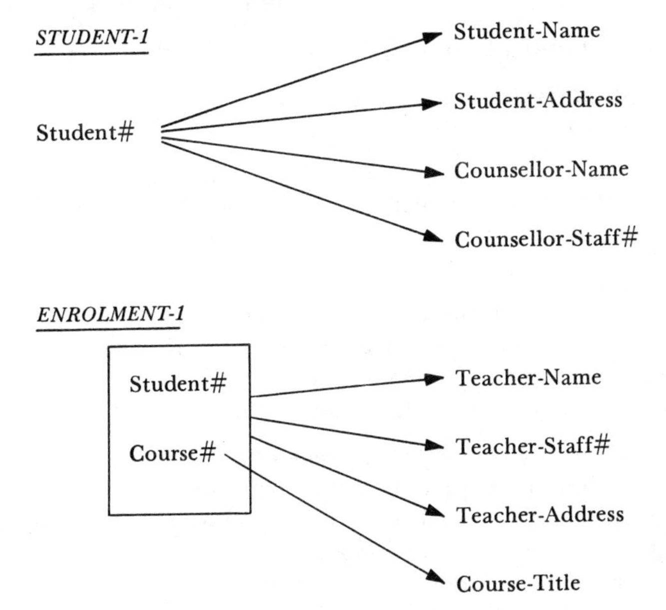

Figure 3.15 *Functional dependencies on identifiers*

This analysis suggests another table COURSE-2, and provides a set of second normal form (2NF) tables. It is a third and a more precise model. The progressive element is the introduction of the new entity set COURSE-2.

STUDENT-2 [Student#: Student-Name, Student-Address, Counsellor-Name, Counsellor-Staff#]

COURSE-2 [Course#: Course-Title]

ENROLMENT-2 [Student#, Course#: Teacher-Name, Teacher-Staff#, Teacher-Address]

3. Investigate the functional dependencies on non-identifiers
The same kinds of problems will arise in the above model as have been discussed, and transitive functional dependencies need to be examined. In this example there are three:

Counsellor-Staff# ——→ Counsellor-Name

and:

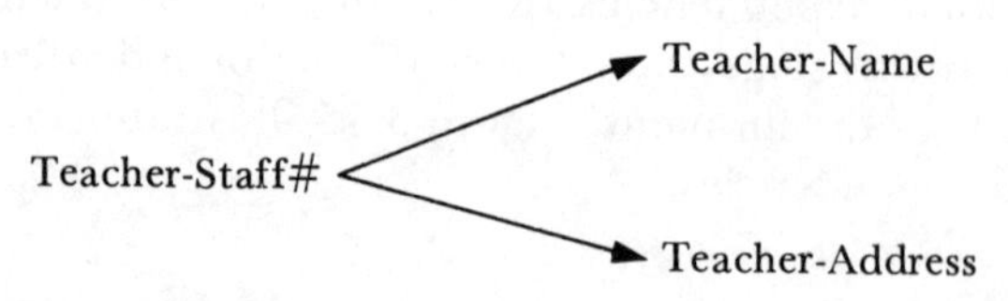

The third normal form (3NF) table model is thus:

STUDENT-3 [Student#: Student-Name, Student-Address, Counsellor-Staff#]
COURSE-3 [Course#: Course-Title]
ENROLMENT-3 [Student#, Course#: Teacher-Staff#]
TEACHER-3 [Teacher-Staff#: Teacher-Name, Teacher-Address]
COUNSELLOR-3 [Counsellor-Staff#: Counsellor-Name]

It is possible, with a certain understanding of the environment, to reach this point in one step. The above tables are now representative of some of the atomic entity sets involved in this task domain. However, there are still some stages that may uncover or suggest the existence of further entity sets.

4. Draw the 'relationships' between entity sets
Relationships are depicted by the one-to-one, one-to-many and many-to-many symbols. This will allow an approximate trans-entity structure to be suggested. This step can be guided by the existence of attribute names that are found in more than one entity set definition. The six possible and independent cases that may be referred to are shown in Figure 3.16.

The diagram is used to suggest the relationships between entity sets. Not all the relationships can be discovered using this diagram (see PARTS and SUPPLIERS in Figure 3.3). The diagram assumes that the shared attribute name in each pair of entity sets means the same, and under this condition the given relationship is a minimum constraint.

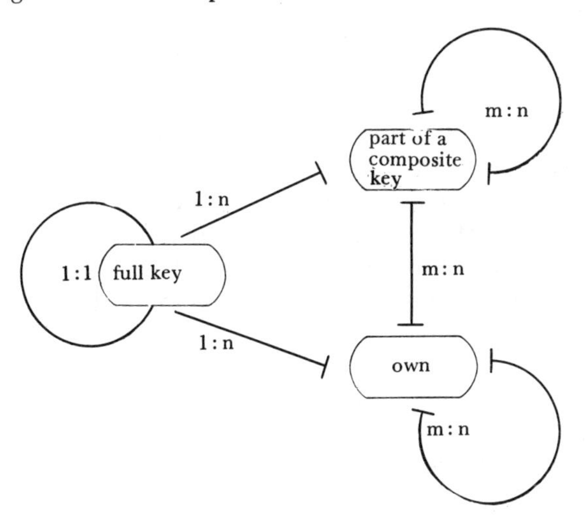

Figure 3.16 *The six cases of relationships between entity sets*

Thus, in the example there is:

ENROLMENT-3 |——— STUDENT-3

because the attribute Student# is part of a composite key in ENROLMENT-3 and is a full key in STUDENT-3. Further, there is:

ENROLMENT-3 |——— TEACHER-3

because Teacher-Staff# is a full key in TEACHER-3 but an own attribute in ENROLMENT-3

5. Expand attributes

Certain attributes may need expanding in order to cope with the given task domain. 'Student-Address' and 'Teacher-Address' can both be expanded to 'Street and Town'. The significant detail here is Town (rather than Street) since it may be useful to know if a teacher and a student come from the same town. Thus, the entity set description becomes:

STUDENT-4 [Student#: Student-Address, Town, Student-Name, Counsellor-Staff#]
TEACHER-4 [Teacher-Staff#: Teacher-Address, Town, Teacher-Name]

The relationship between these entity sets is:

STUDENT-4 |———| TEACHER-4

because Town is now an own attribute in both entity sets (see Figure 3.16).

6. Create new entity sets to reduce n:m to 1:n
The many-to-many relationship usually implies the existence of an entity set, and in the above example it is the entity set TOWN-5. Figure 3.17 shows this new entity set. However, a new entity set may need to be defined that has a key constructed from the combination of the keys of the component entity sets (eg [Student#, Teacher-Staff#:]).

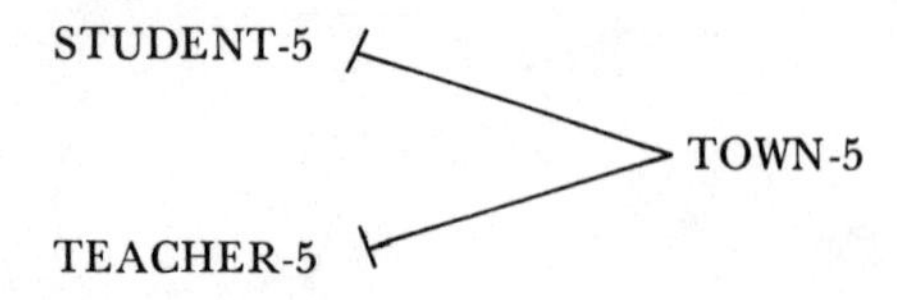

Figure 3.17 *TOWN-5 as implied by the many-to-many relationship*

7. Replace all established 1:n relationships with 0-implies or 1-implies
This will require further interrogation of the client and examination of the environmental data to fix the mapping component.

8. Look for join dependencies
This stage is equivalent to determination of fourth and fifth normal forms. However, the multivalued dependencies and join dependencies are only a concern of the key attributes. The result for this example will be similar to Figure 3.7.

9. Consider entity restrictions
Entity restrictions usually occur where there is an entity set defining a directed graph of some kind. These will be such entity sets as 'parent-child' where a parent cannot be his or her own child.

10. Determine partitions
Partitions will make the definition more concise and suggest

different roles entities may play. Thus a student, counsellor and teacher are all people who live somewhere, and there emerges another entity set PERSON. Further, counsellors and teachers are members of staff. All this information can be depicted in one single Sfd graph (see Figure 3.18).

The determination of restrictions, partitions and other constraints in each case is non-progressive in that new entity sets are usually not produced. All these constraints simply restrict what is already established.

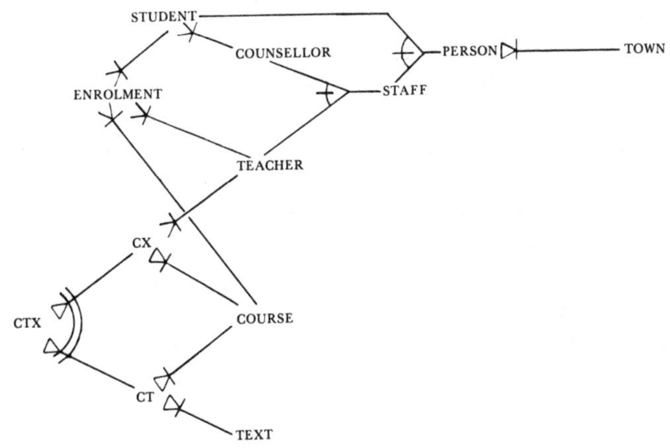

Figure 3.18 *The Sfd graph of the student-course environment*

The formation of this conceptual model has gone through ten stages where each stage attempts to make a better specification than the last. The kinds of imprecision likely to occur have been discussed at length in Chapter 2 and in earlier sections of this chapter. Each stage has an associated set of questions that needs to be answered by the client and should be supported by examples. The models produced at each stage can be checked for consistency, and each model represents a new theory to be investigated. The final Sfd graph may need redrawing many times (eg six or seven times) before a satisfactory homomorphism is established between it and the client's statements. Other details may be captured through cardinality constraints and update restrictions (these are auxiliary hypotheses). There are, however, elements of the task domain that cannot be expressed in ERA, and some of these will be mentioned in Chapter 5.

Conclusions

Since the term database was coined there has been an important shift of emphasis. Initially it was simply a question of data independence that was to be achieved. There was no reason to believe that it was possible, only that it was desirable. Combine this desire with the needs for data protection, methods for rational design and a 'natural' interface to a database retrieval system, then the conceptual database follows. The route from files to conceptual models has been through the analysis derived from relations which attempts to isolate the atomic concepts that make up the information needed for a task domain. The relational model could not be used without reference to the table and the simple computer file models. These still needed extending to capture the essential semantic character of fact representation. To this end the conceptual model required an entity set to be represented by at least a BCNF table. The proposed semantic function captures the requisite granularity of the task domain and the essence of the perceived environmental constraints.

Many of the important characteristics of the environment are contained within the conceptual model framework. This analysis focuses on defining the types of constraints that can be imposed on the information of a task domain at a particular level of abstraction. The objective is to lay down the machinery of the environment so abstracted as to expose its gears and describe how it meshes. This can be done only through interviews with the client (users). The semantic nature of the conceptual model makes it an ideal tool for discourse control as the client and analyst create a description to fit the client's perceptions. Once this knowledge of the environment is obtained then the skeleton for the contextual representation of knowledge (Sfd graph) is the basis for system design. This representation is supported by a machine, and the machine is another source of constraints of which the users, at this stage, ought not to be aware. The underlying computer file model that is presumed is far too simple for practical systems, and better-engineered designs of greater complexity need to be explored (see Chapter 4).

The Sfd graph has evolved to satisfy the requirements of design, elicitation, clarity and precision. The phenomological stance initially adopted was in order to shape an approach that is appropriate to the task of design. It is a powerful tool to be used for developing knowledge-based systems, and it provides a framework upon which models of reason may rest.

CHAPTER 4

Engineering Conceptual Databases

> Still round the corner there may wait
> A new road or a secret gate;
> And though I oft have passed them by,
> A day will come at last when I
> Shall take the hidden paths that run
> West of the Moon, East of the Sun.
>
> *Frodo's song*
> *The Lord of the Rings by J.R.R. Tolkien (1954)*
> *George Allen and Unwin Ltd (1954)*

> Your only chance now is to catch our
> brother the North Wind. He is the oldest
> and strongest of us all, and the only places
> he has never blown on are the places that
> aren't there.
>
> *East of the Sun and West of the Moon*
> *from Scandinavian Legends by Gwyn Jones*
> *Oxford University Press (1956)*

Introduction

Knowledge representation schemes are not arbitrary; they are expected to be a step towards symbol manipulation systems and as such they will be influenced by the machines on which they may be given concrete form. The language chosen to describe the knowledge should enable the knowledge structure to be made explicit in a way that will eventually lead to an efficient design. However, if the representation scheme is tied too closely to the machine, then there is a danger that the description will become prejudiced by physical properties that have little to do with the knowledge domain. For this reason, it is necessary to ensure that the initial analysis is concerned with a knowledge description that keeps the underlying engineering requirements subdued. However, there needs to be another representation scheme that emphasizes the machine architecture in a way that makes clear the features that relate to its performance, each scheme illustrating different characteristics for implementation. The two (or more) sequences of representation schemes should fit naturally together and provide a medium for progressive and coherent design. Figure 4.1 suggests a generalization of the simple computer model (that influenced the arguments that formed the table model) to include other kinds of computational machinery. The construction of the conceptual database is determined by the type of machine architecture employed in its realization.

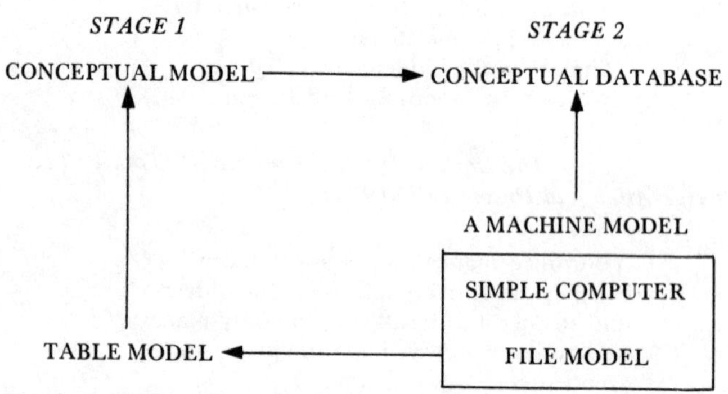

Figure 4.1 *Two stages in engineering a knowledge-based system*

There is another more important issue concerning the dynamics of conceptual database development. Studies have shown (Belady and Lehman, 1976; Lehman, 1980) that computer systems designed to serve a purpose within an organization are subject to recurrent re-specification. This re-specification is needed because the initial requirements are not properly identified and because of the normal development experienced by any living group. The result is frequent modifications to the programs and file structures with the subsequent deterioration in performance and increase in errors. It has been shown (Torsun, 1981) that in mature programs much of the code (about 30%) lies idle because it cannot logically be entered from any normal use of the program. The source of the problem resides in the incoherent approaches to program construction, approaches that do not acknowledge the epistemic aspects.

Logical models and schemas

The logical model is a term used to describe a particular conceptual database using an accepted machine model. It is also a programmer's perception of data structures in terms that are related closely to data storage and access paths (the physical model). In general, the logical model may not be representative of the actual storage in a particular case. However, it should be expressed in a manner that suggests potential storage structures on the machine so that all the essential characteristics of data access are apparent. A particular logical model is referred to as a schema when the schema reflects all the current storage structures. A sub-schema is another logical model derived from the schema that indicates an alternative or partial organization of all, or part, of the original data.

The initial purpose of the schemas and sub-schemas is to provide an interface to a database that can be shared by many different programs. A sub-schema is a way of selecting an interface for a particular group of people that have a common set of tasks. The schema is intended to be a consensus of each group's needs for a data conglomerate. It is a consensus based upon machine behaviour as well as human usage. The database manager has to choose a storage organization that results in a satisfactory overall (or average) performance. Consequently, certain users (and maybe all) will get less than optimum service from the central database.

The primitive elements of the machine model used in logical models are structured records and pointers. Records can be stored together in files and linked together via the pointers. It is presumed that only a single record at a time is available to a programmer for processing, and the next record can only be retrieved by either 'next in a file' or 'following a pointer'. This machine model has evolved from machine architectures based upon magnetic tape storage, mechanical card sorters and the von Neumann machine. Behind these primitives there is further information concerning access methods. These include sequential and inverted indexing. However, new machines require different kinds of models to characterize their storage and access methods (eg the Content Addressable File Store, the Distributed Array Processor and the Dataflow Computer). To design a logical model requires a good understanding of the mechanism, and many design decisions need to be based upon unstated features of the machine architecture.

RELATIONAL ALGEBRA

The simplest design mapping from a conceptual to a logical model is to use a machine model such that every table is a file and every row in a table is a record indexed on the primary key attribute. The machine model will have relational operators to manipulate the tables. This machine model is the simple computer file model initially presumed for relational analysis. In the following description the 'computer file' will be referred to as the 'table' and the 'relational operations' will be the system functions available to a programmer.

There are two types of relational operations which are quite distinct. The first type refers to set operations and in certain cases can only be performed on tables that are union-compatible. Tables are union-compatible if they are of the same degree (same number of attributes) and the corresponding attributes of the two tables are drawn from the same domain. The following shows two union-compatible tables and the associated operations. The method of presentation of the relational operations is based upon the IBM System R (Date, 1981) user interface. The lexigraphic symbols shown in parentheses are an alternative representation. The example tables are:

PART[Part#:	Description,	Price]
1	nut	6
3	bolt	3
6	bolt	10

COMPONENT[Comp#:	Name,	Price]
5	plate	5
4	angle	20
3	bolt	3

where Part# and Comp# are attributes drawn from the same domain of numbers and Description and Name are attributes drawn from the same domain of character strings. The assumption is that entity Part# 3 is the same entity as Comp# 3. The possible set operations on these tables are:

1. *UNION* (∪)
   (PART ∪ COMPONENT)
   PART *UNION* COMPONENT

1	nut	6
3	bolt	3
6	bolt	10
5	plate	5
4	angle	20

2. *INTERSECTION* (∩)
   (PART ∩ COMPONENT)
   PART *INTERSECT* COMPONENT

   3 bolt 3

3. *MINUS* (−)
   (PART-COMPONENT)
   PART *MINUS* COMPONENT

   1 nut 6
   6 bolt 10

   The cross product (TIMES or x) of two tables does not have to be union-compatible and is used to define the JOIN (*).

The second type are the relational operations, and these are:

4. *SELECTION* (|⟨Boolean⟩)
   (PART|*Description = 'bolt')*
   PART *WHERE Description = 'bolt'*

   3 bolt 3
   6 bolt 10

5. *PROJECTION* (Π)
   (Π Description PART)
   PART */*Description*/*

   nut
   bolt

6. *JOIN* (*)
   Given:

A[a, b:]		B[b, c:]	
1	x	x	A
2	x	x	B
2	y	y	A

   Then:
   (A*B)
   A *JOIN* B

[a, b, c:]		
1	x	A
1	x	B
2	x	A
2	x	B
2	y	A

The natural join is simply the cross product (*TIMES*) of all those tuples that have equal values within a common attribute. This suggests that the join could be implemented through a *SORT* on the common attribute followed by a cross product between the records on each file that share the same value for that attribute. Thus, the new (unnamed) table above could be formed by taking the *UNION* of the two sets:

{(1, x)(2, x)} *TIMES* {(x, A)(x, B)}

and:

{(2, y)} *TIMES* {(y, A)}

and mapping the common attribute name under one heading.

Sometimes a *JOIN* is required over attributes of different names (but compatible domains). An example is where GRANDPARENTS are derived from the relation PARENT [ParentName, Child:] by a self-join. This requires the relation PARENT to be joined to itself such that PARENT1.Child is equated with PARENT2.ParentName. This may be done through the natural join as before by renaming the attributes in PARENT1 (say). However, this kind of join will be expressed through listing, in order, each of the attribute names for each relation. The first in the list of relation 1 will be compared with the first in the list of relation 2. The two lists will be separated by the operator *JOIN*, the first list preceded by a dot (.) and the second terminated by a dot (.). So that:

RELATION1.⟨list of attributes⟩ *JOIN* ⟨list of attributes⟩.RELATION2

and in this example (where:= means 'become equal to'):

GRANDPARENT := PARENT.Child *JOIN* ParentName.PARENT

The join symbol * may also be used in place of *JOIN*, and .*. with empty join lists is interpreted as a cross product.

There are many variations on this theme (non-natural joins) that will not be described in detail here. One example is the cross product of common attributes where one value is greater than the other. In this case it is not usual to combine the attribute names, although it could be done if there is either a procedure for choosing the value to be kept (ie the longest value) or there is a domain of pairs created. However, not all the desirable manipulations can be achieved through relational algebra as presented. For example, in the case where a table is to be generated that describes when certain processes can be carried out in the machine-shop, from the information in Figure 4.2, then non-relational operations are required. Note that this example also illustrates the extra information carried by the MACHINESTAFF relation (ie who operates what and to what level of skill).

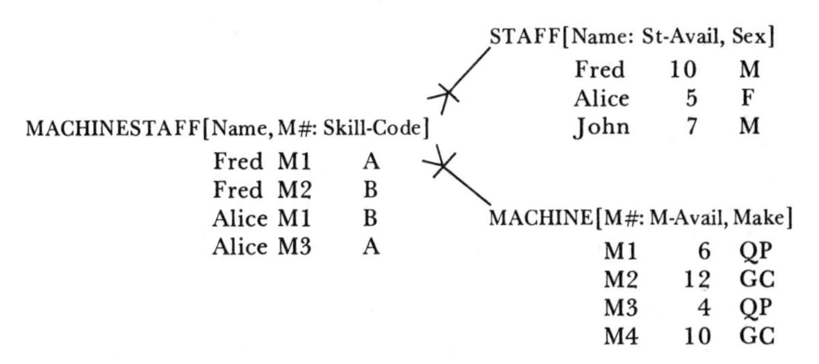

Figure 4.2 *The machine-shop*

The machine-shop data give details of when (given as week number, say) staff become available (St-Avail) and when machines become available (M-Avail). A process cannot be carried out until the skilled staff that can run a machine and the machine are both available. This requires linking the appropriate staff to the machines they can operate, and comparing the two availability times.

The semantic functional dependency (Sfd) graph shows that MACHINESTAFF *JOIN* STAFF over the common attribute 'Name' will provide the link between staff availability and the machines. It is through the join that MACHINESTAFF inherits the associated properties of STAFF. Similarly, a join between MACHINESTAFF and MACHINE will provide a link between machine availability and staff. All the required information is found in the table resulting from the operations:

(MACHINESTAFF *JOIN* STAFF) *JOIN* MACHINE

However, what is required is the *selection* of the latest availability time from the staff and the machine for *each case* of MACHINESTAFF so that we may obtain the following:

[Name,	M#,	GRTST,	Skill,	Sex,	Make]
Fred	M1	10	A	M	QP
Alice	M1	6	B	F	QP
Fred	M2	12	B	M	GC
Alice	M3	5	A	F	QP

This same result can be achieved through the natural join with an enhanced selection function GRTST used for each value of M#. This cannot be achieved simply through the normal relational operations because the function GRTST has to be limited at each stage by the value of M#. In practice, as a system function, this application of GRTST is best done *during* the *JOIN*.

In order that the set of operations is relationally complete (see p. 35), there is the process of division (÷). Very roughly, A *DIVIDEBY* B means 'return the attribute (columns) *not* in relation B for only those rows in the relation A that have every value of the attributes given in B'. For example, if we needed to know who can operate the machine M1, then a table B can be created:

B[M#:]
M1

and:

MACHINESTAFF *DIVIDEBY* B

will return:

[Name:	Skill-code]
Fred	A
Alice	B

Alternatively, if it was required to know who can operate *both* M1 and M3, then a table B′ can be created:

B′[M#:]
M1
M3

and:

MACHINESTAFF[Name, M#] *DIVIDEBY* B′

will return:

[Name:]
Alice

Note that this questions responds with a table made up of an attribute that is the difference between the dividend and the divisor.

The algebra can now be used to create new tables, and these tables will be meaningful provided every operation is done with reference to the Sfd graph. For example, consider the following queries about the machine-shop (see Figure 4.2).

1. Get machine.make for machines that are operated by Alice:

   ((MACHINESTAFF *JOIN* MACHINE) *WHERE* Name = 'Alice') [Make]
   will return:
   [Make:]
   QP

2. Get machines for machines operated by at least one male:

   (((STAFF *WHERE* sex = M) [Name]) *JOIN* MACHINESTAFF) [M#]
   will return:
   [M#:]
   M1
   M2

3. Get machines that are operated by both sexes:

   (MACHINESTAFF *JOIN* STAFF [Sex, M#]) *DIVIDEBY*(STAFF [Sex])
   will return:
   [M#:]
   M1

There is a process of amelioration (to improve performance) that can automatically be applied to these queries (Smith and Chang, 1975; Hall, 1976). Amelioration transforms the queries so that the same result is achieved with minimum processing. The principle depends upon reducing the amount of data to be handled by ensuring that selections occur before joins and that selections are combined so that a relation need only be scanned once. Different kinds of machines have different considerations.

HIERARCHICAL STORAGE STRUCTURES

The original purchasing system described in Chapter 3 (see Figure 3.5) held the data on magnetic tape. The records were hierarchical and were indexed on Part#. Each directed path or subpath of the Sfd graph can be translated into a hierarchical storage structure. For convenience the Sfd graph in Figure 3.5 is reproduced in Figure 4.3.

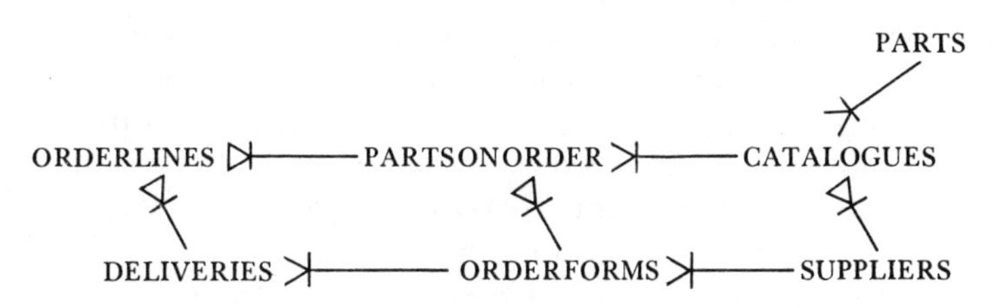

Figure 4.3 *Sfd graph for a purchasing database (see Figure 3.5)*

The 0-implies and 1-implies both represent a many-to-one relationship, thus every possible path through the Sfd graph starting at the head (PARTS or SUPPLIERS) represents a potential hierarchical structure. In this case, all the structures derived from complete paths are shown in Figure 4.4 together with example records in Figure 4.5. The component relationships of the 0-implies and 1-implies shown by the symbols described in Chapter 3 can be used on their own. Each path can be decomposed into the paths of the single components. These symbols then show the one-to-one, many-to-one and many-to-many relationships between entity set key attributes (see Figure 4.4). The 'onto' component can be retained as a simple triangle marking an arc.

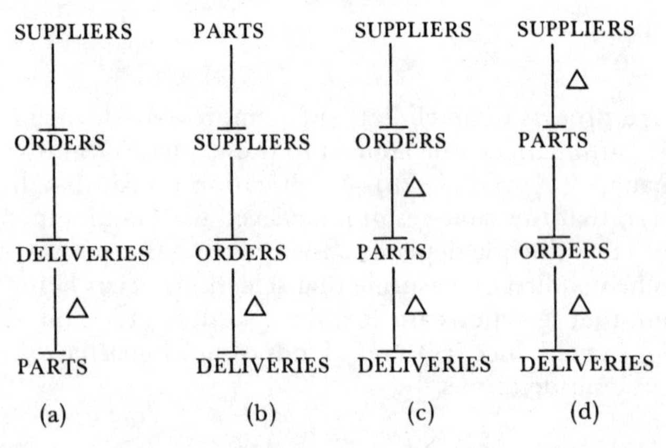

Figure 4.4 *The many-to-one relationship between key attributes (levels)*

The von Neumann architecture defines a linear storage of data and program. This linear storage imposes certain constraints. In order that data can be related the data items can either exist in consecutive storage areas or can be addressed by pointers stored as physically related data items. Multiple groupings of related data items such as the many orders placed on a particular supplier can be kept together as a 'repeating group' within a 'record'. A *record* is a structured storage area that can be retrieved or accessed so that all the data is immediately available to a program for processing. Sometimes it is desirable to store pointers to other records of a different type that contain related data. For example, a record that has information about suppliers such as address, telephone number etc, may refer to the parts sold by that supplier as a repeating group of pointers to records containing details of the parts. The parts records, in turn, may refer to the suppliers that supply that part. There are many ways in which data may be linked through access paths, and it is normal practice to make the access paths reflect the same constraints as expressed by the relationships between entity sets. However, it may be that engineering requirements enforce a mismatch between the physical, logical and the conceptual.

An example of a linear record structure would be a sequence of storage elements. A storage element consists of a series of closely related attribute values laid down in a fixed pattern. Thus in the S element supplier 'number' would be immediately followed by supplier 'name', and each attribute would be in a predefined length of store. All 'names' would be of fixed length. The records illustrated in Figure 4.5 would consist of the following sequence of storage elements:

(a) S1 O2 D2 P1 O1 D1 P1 D3 P2 D5 P2

(b) P1 S1 O1 D1 O2 D2 S2 O3 D3 D5

(c) S1 O2 P1 D2 O1 P1 D1 P2 D3 D5

(d) S1 P1 O1 D1 O2 D2 P2 O1 D3 D5

Associated with each storage element is a relative pointer that addresses (relative to current address) the next element at the same level of the hierarchy. The top level (eg S1 in a) addresses (absolutely) the next record in the file, the second level (eg 02 in a) has a relative pointer to the next element at that level (eg 01 in a). It is more compact and faster to process records of identical structure where relative addressing is not needed.

In order to distinguish between the physical, logical and the conceptual specifications a new set of symbols is defined. These symbols give the logical view of the physical layout of the data. For physically consecutive data items that form a record a 'hard' arrow is used, and for pointers between records a 'dashed' arrow is used thus:

A ⟷ B Data items of type A and B are in a one-to-one relationship. A and B usually refer to sets of attributes identified by their key attribute.

A◄——►►B B represents a repeating group of data items in a record denoted by A. It is possible to chain these arrows together to form a repeating group within a repeating group. A is the principal record data item, and B is the secondary item and so on. The individual groups of B may be referenced by relative pointers within the record.

A - - > B A record of type A has a single pointer to a record of type B.

A < - > B A record of type A has a single pointer to a record of type B, and each record of type B points back to that record of type A.

A - - >> B There is a repeating group of pointers in the record of type A pointing to records of type B.

A << >> B There are repeating groups of cross-referencing pointers in both record types A and B.

The rule is that repeating groups can always have zero or more members, but for physically consecutive data items there must always be a principal record data item. Each data item relies upon the existence of the next item up in the hierarchy. Figure 4.5 gives some example records that would be formed from the logical view given in Figure 4.4.

Records of the same type form collections called files that are grouped either because the records are physically together in a defined area of store or because they are linked through a series of pointers. Files may also be defined by a file of indexes that addresses a member record. File pointer sequences are usually not illustrated with the above symbols.

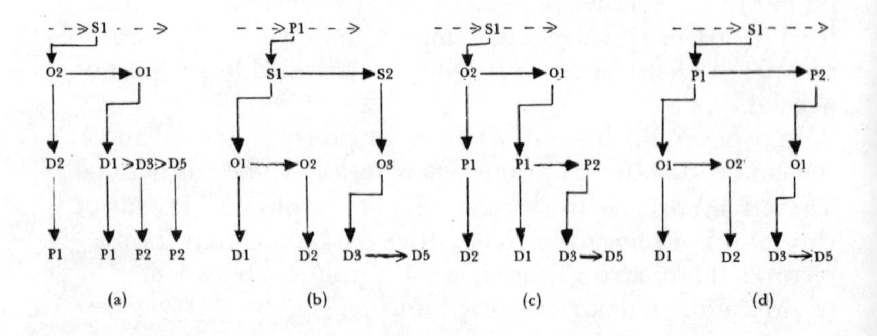

Figure 4.5 *Example records (keys only given) corresponding to the many-to-one relationships (a physical model)*

Any one of the hierarchical structures illustrated in Figure 4.4 could have been chosen for the purchasing database provided the Sfd graph consisted of only 1-implies. For example (see Figure 4.4) if (a), (c) or (d) were chosen then information about a Part# could not be sustained unless it had in (a) a delivery, in (c) an order for it and in (d) a supplier. The 1-implies between CATALOGUES and SUPPLIERS (see d) suggests that the original choice (b) was in fact the best (given that a single data structure must be chosen). Under these conditions certain queries are very easy to answer. For example:

1. Get supplier addresses for Part# = 1247.
2. Get deliveries of Part# = 1247 from Sup# = 7132.

However, there are other kinds of questions that demand a considerable amount of processing such as:

3. Get all the Part# on Order# = S54321.
4. Get every Part# that was delivered in 1983 from Sup# = 7132.

Answers to these questions are more easily obtained through the other structures.

In order to maintain the same constraints and the implied performance for queries suggested by the Sfd graph (see Figure 4.3), then every structure shown (see Figure 4.4) would have to be kept and maintained. This results in update problems. An alternative approach would be to construct a set of files through a process of controlled 'denormalization' (see Figure 4.6). Denormalization means the re-creation of a structured representation from the 'normalized' entity sets; the mappings are not exact between the constraints and the storage relationships. This will introduce constraints in the system that will increase the difficulty of certain tasks, and, in some cases, prevent tasks from being done.

The three files given in Figure 4.6 provide the same constraints as suggested by the Sfd graph and yet maximize the amount of pre-compiled information (consecutive storage date items). Further, this pre-compiled information (which is the effect of the hierarchical structure) is in a form that fits well the set of tasks to be performed by the purchasing department. Note how the 1-implies link is used. The 1-implies demands that an implied entity cannot exist without at least one other entity implying it. Choosing the 0-implies link, in the case of (c), results in the complementary 1-implies

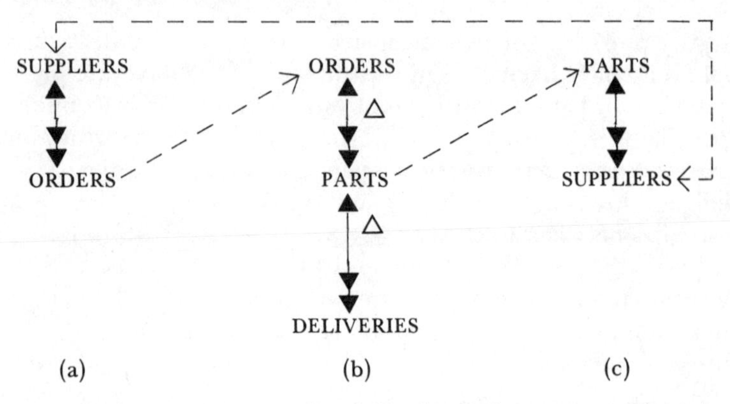

Figure 4.6 *A set of 'denormalized' files (dashed arrows are pointers)*

dependency being maintained automatically through a physical dependency (see Figure 4.4(b) and (d)). In Figure 4.6(c) a part can exist without a supplier but not the other way round. This is exactly the requirement specified in the Sfd graph. If the choice had been taken from SUPPLIERS 1-implies PARTS, a degree of freedom would have been lost and an extra unwanted degree of freedom between ORDERS and PARTS gained. However, in (b) PARTS have a different function (see below), and the restriction is required to reflect this new meaning.

In order to avoid scanning files record by record, access pointers in (a), (b) and (c) can be used. The links are between ORDERS (in a and b), PARTS (in b and c) and SUPPLIERS (in a and c). To avoid repetition and to ensure consistency, own attributes are kept in records that are the main index of a file. Note that, since Part# can already be stored independently in (c), there is good reason to place ORDERS as the main index for file (b) despite the 1-implies between ORDERFORMS and PARTSONORDER. PARTS in (c) are parts in catalogues; PARTS in (b) are those on order. This is exactly the situation described.

RING-STRUCTURED STORAGE

Magnetic disk introduced new possibilities for storage structures and the CODASYL (Conference on Data Systems Languages) committee (1971) was formed to define new commercial standards. COBOL was the principal commercial language that had been designed for electronic computers primarily to process data from cards during a time when the most successful machine was the electro-mechanical Hollerith card sorter. The limit of 80 characters per line in the original design of the cards was even carried through to visual display units and influenced the style of many program editors developed 60 years later. A version of this machine was used by the United States Census Bureau in 1890 to keep track of immigrants, and its offspring was used during the Second World War to select United States military staff for special missions (Rogers, 1969). The CODASYL committee was faced with the problem of extending the ideas of COBOL rather than starting with revolutionary new concepts. This was necessary because of the vast amount of investment in COBOL. The result was a new definition of 'next record' to take into account a record's position within a network of pointers. The networks were restricted to those based upon ring structures.

The ring structure principle represented an enhanced machine model which was derived from list processing (LISP) data storage management methods (McCarthy *et al.*, 1965). The methods had been based upon Lambda calculus, and they had proved to be very effective as a generalized symbol manipulation mechanism (this connection between the calculus and design methodology was not retained). However, the method had to be modified to operate outside main memory and on a disk in a way that was coherent with the COBOL principles. The result was that files became lists of records, all of the same type, where each record contains a pointer to the next record in a file. The last record, instead of containing no 'next' pointer, points back to the first record of the list (head). The purpose behind this ring was to reduce the potential for termination errors by standardizing the format of all records.

In this structured system it is possible to link the rings together by providing further pointers. These linkages are controlled by the introduction of set (set members) and set owners. It is through a set that a set owner can be found, that in turn may belong to a set. The linkage between different types of records is called a set type and is strongly associated with a 'relationship'. Thus, tree structures could be represented as well as lists. These tree structures are homomorphic to repeated groups (eg see Figure 4.6) as originally required for tape-based systems. There is a bonus in that attribute values need only be stored once since any reference to them can be done via a pointer. In addition, this provided consistency of value for all records referring to the same item in store.

Ring-structured storage was also used as a representation scheme. The Bachman diagrams, similar to those illustrations given in Figures 4.4 and 4.6, provide a method of sketching a particular storage structure of this kind and are heavily biased towards implementation. This distorts the analysis by introducing unwanted constraints, and imposes a foreign perception of entity sets. Individual occurrences of records can be expressed but the one-to-many (and one-to-one) symbols (as used in Figures 4.4 and 4.6) eliminated the need to be explicit at that level of detail. However, the actual implementation may be more complex than these diagrams suggest because there may be (for example) one- or two-way pointers linking the records.

This representation scheme does not provide any guidance in systems design since there is no test to determine the 'correctness' of the entities being represented nor the restrictions on the operations; there is no definite connection with a firm theory of data constructs. One simple error is the 'connection trap' where wrong conclusions are drawn from linkages. For example, consider a situation where there are files of PARTS, JOBS and SUPPLIERS (see p. 64 and Figure 3.8). It would seem reasonable that if a particular job uses some parts, then a pointer to these parts should be sufficient to indicate this relationship. Further if these parts are being supplied by a particular supplier, then again it would seem appropriate to simply point to the supplier. This situation is illustrated in Figure 4.7 where Job J1 uses parts P1 and P2 and job J2 uses parts P2 and P3. Supplier S1 supplies P1 and P2, and supplier S2 supplies P2 and P3. The equivalent general description is given in Figure 4.7(b).

The conclusion that may be drawn from these linkages is that job J1 may have used part P2 from supplier S2 when in fact the only supplier used by J1 was S1. There is no way in which this information can be represented or recovered from this structure. There is also the problem of where to store information about the relative prices of parts from different suppliers. Is part P2 cheaper from supplier S2 than from S1?

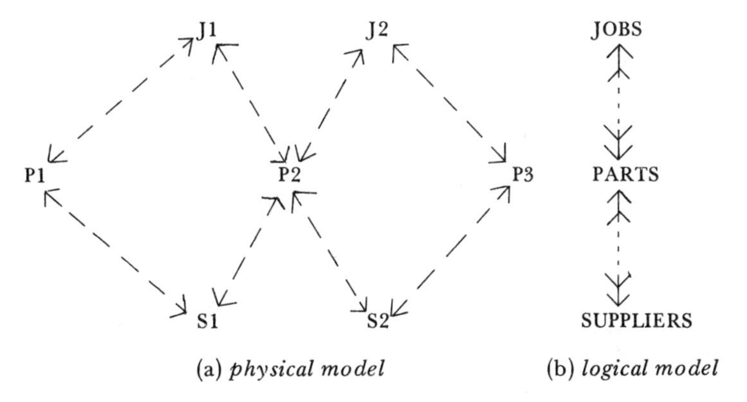

(a) *physical model* (b) *logical model*

Figure 4.7 *An illustration of the connection trap*

Does Job J2 get P2 from both suppliers and in what ratio? Where should the quantity of parts used on a job be kept?

The above problems are not solved by representation alone. An Sfd graph and the associated tables must be constructed to clarify the entity sets employed in the task domain. The entity sets have to be drawn out from the users. They are not immediately apparent. The protocol should reduce the possibility of errors occurring by laying bare the components of the required system from the users' point of view. The transformation between Sfd graphs and ring-structured storage is particularly straightforward. Each table becomes a file (ring), and each implication becomes a pointer from set members to set owner. Figure 4.8 (a and b) illustrates this transformation for the purchasing database (for comparison purposes the same values and relationships have been used as in Figure 4.5).

The symbols at each node represent records that contain own attribute values, as well as either pointers to records containing the key attribute values (in which case they are called 'linker' records) or the key attribute values themselves. The physical constraints imposed by the structure are used to reflect the constraints expressed in the Sfd graph. In Figure 4.8 the constraint PARTSONORDER $\rightarrowtail\!\!\!+\!\!\!-$ CATALOGUES can be maintained without special checks by PARTSONORDER records with pointers to part values via the CATALOGUES linker records.

The important characteristic of this structure is that every record of the type (ring) has exactly the same number of

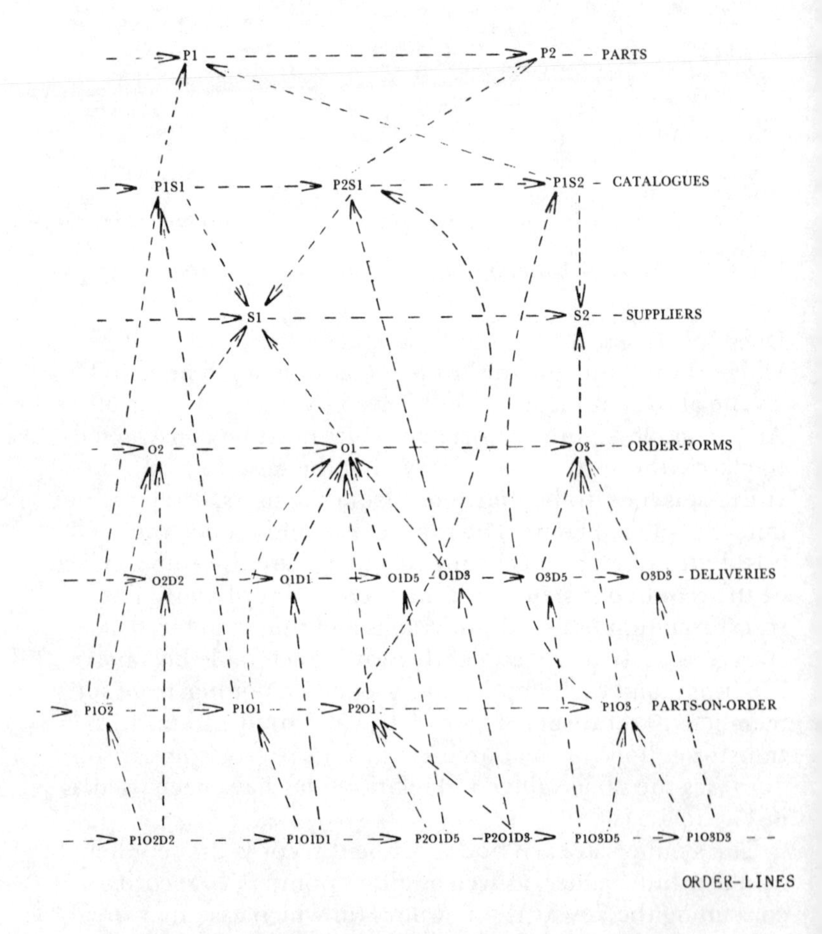

Figure 4.8(a) *The physical model of the ring-structured example of the purchasing database*

pointers going outwards. Thus, routes through the system can be defined very simply and efficiently in software. The problems with this structure are that much space is occupied by pointers, and time is wasted moving the heads of the disk to recover conceptual neighbours that are physically distant. Further, the difficulty in maintaining this structure during

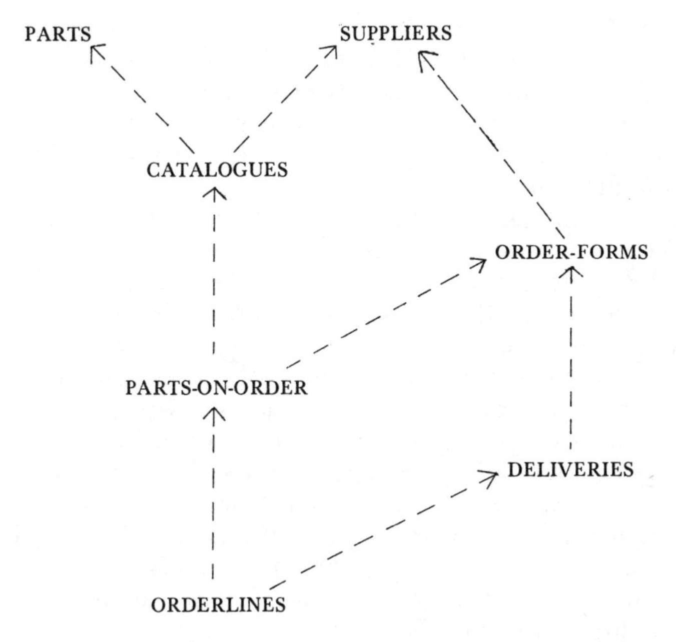

Figure 4.8(b) *The logical model of the purchasing database*

the insertion and deletion of data is considerable, and (in time) the performance of the system degrades as related information is scattered throughout the disk surfaces. More problems arise when new entity sets are introduced, new relationships defined and old concepts modified.

Storage structures on the Content Addressable File Store (CAFS)

INTRODUCTION

The difficulties of inserting and retrieving information from a storage medium arise mainly because of the need to define a structure in which the information can reside. Once such a structure is defined and the information placed within it there follows the complementary difficulty of finding the information. Information is nearly always required because it conforms to some desired pattern (see Chapter 8) and *not* because it is stored in some particular place. It would resolve many difficulties and make considerable simplifications if information was stored and retrieved in a machine that was addressable according to the *pattern* of the information. It was for this very fundamental reason that content addressable memories have been constructed and, in particular, the ICL Content Addressable File Store (CAFS).

This section will describe first the experimental MKI CAFS because it contained many good ideas that were shelved for reasons of commercial acceptability rather than technical considerations, and second the CAFS 800 (MKII) because it is commercially available. Other CAFS engines are either under development or already available (eg CAFS ISP). The description of the CAFS is necessary because it introduces new design criteria with which most people will not be familiar. The CAFS resolves in one single move many of the difficulties encountered with the traditional storage systems.

The history of the CAFS dates back prior to 1970 when the original purpose in the design of the ICL CAFS was to provide a total inverted indexing machine to a database (Coulouris *et al.*, 1972; Mitchell, 1976). Each CAFS record was conceived to be either a string of keywords or a list of attribute/value pairs to represent a document summary. The actual document was expected to be stored on a separate disk or disks that would be addressed and retrieved in the normal way. This was to be called the Direct Access Store (DAS) but it was never constructed. This initial design envisaged that the CAFS (index) records, stored on disk (of capacity 8, 30 or 60 Mchr) would be continually and exhaustively scanned by a hardware matching system. The patterns to be matched and the associated information to be retrieved would be set up in the Disk Controller Unit (DCU) so that the processing occurred separately from the central processor. The DCU contained small memories (that held the pattern elements) called key registers along with a simple, but fast, special purpose processor (the search evaluation unit). A pattern element was a simple attribute/value pair, and a pattern consisted of a Boolean statement. This processor would compare and calculate the pattern to be matched with each CAFS record as it passed under the disk heads along with its associated document address in the DAS. If there was a 'hit' (the CAFS record matched the elements and satisfied the criteria specified), then the document would be retrieved from the DAS. Tasks (requests for documents) could be continually added to the processor, and since the disk was exhaustively scanned there would be no need to delay the other matching activities.

The CAFS processor in the disk controller was capable of handling much more data than a single disk head could provide in a given time, and thus each of the ten heads of the disk were used in parallel. This required each head to have its own amplifier instead of a single amplifier switched from head to head. The initial experimental CAFS (CAFS MK1) used an ICL EDS 8 disk (and later extended to 30 Mchr) with the records stored vertically and offset on each plane. The records in sequence thus formed a spiral pattern through each of the ten surfaces of the disk and restarted again on the top surface. However, this pattern of physical storage was never used in the final products because of problems of compatibility with existing storage methods.

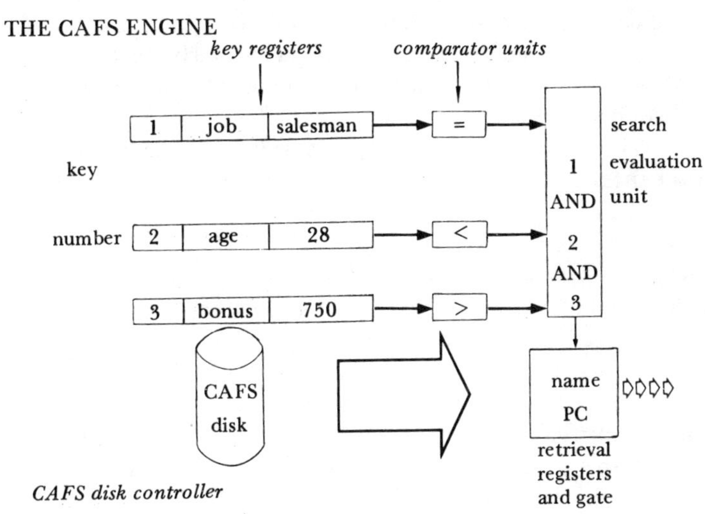

Figure 4.9(a) *A simple schematic diagram of the ICL CAFS*

REQUEST

GET Name, PC *FOR* Job = Salesman
AND Age < 28
AND Bonus > 750

RESULT:

Name	PC
Brown	77321
Smith	4413
Smith	4562
Jones	8231

Figure 4.9(b) *A simple example of an ICL CAFS task*

A view evolved of the CAFS as a database engine in its own right rather than an adjunct to a database system. Figure 4.9(a) illustrates the main features of the CAFS architecture that were carried forward from the initial perception of the machine. The key registers are set to contain field/value pairs (a pattern element) to be matched against each record that passes under the heads. A CAFS record is a sequence of fields containing an identifier, length and value. The record terminates in a special field indicated by a special field identifier.

In the initial design there were 24 short key registers (120 characters each), but later designs greatly increased the size of each key register so that lengths of text could be matched. Areas of the key registers can be 'masked' so that only a part of a pattern element (a value) is detected. Any pattern of masking can be defined so that individual bits or characters can be ignored in an element. This is particularly useful for English word stem matching so that elements in a CAFS record such as 'big, bigger, biggest' can all appear as just 'big' to the CAFS controller. Masking of this kind is indicated at the user interface as a back arrow (←) thus:

big ←

The key comparators units are a four-bit latch. These four bits represent =, <, >, ≠ for each key register. They are all set to 0 at the beginning of each record unless specially locked on. It is thus possible to have several fields of the same identifier in a record with different values.

The search evaluation unit can be fully exploited by expressing many complex Boolean and non-Boolean functions. Simple expressions such as the request:

GET name, personnel-code *FOR* Job = Salesman
AND Age < 28
AND Bonus > 750

may be expanded to include not only many levels of brackets, absence of fields and 'nots', but also field comparisons (ie Age > wifes-age) and the quorum function.

The quorum function is a function that takes any list of M Boolean expressions and returns 'True' if N out of the M expressions are 'True' where $N \leqslant M$. This is equivalent to assigning the number 1 to 'True' and 0 to 'False' and adding up the list of M numbers. If $M \geqslant n$, then the list returns 1 (ie 'True'). The full generalization of the quorum function allows each term or expression (eg Job = Salesman) to have an assigned weight other than 1 if True. The weights of each term or expression successfully matched are added together, and if the final total is greater than a certain given value, then the retrieval registers and gate allow selected information through to the central processor unit. Several refinements have been added to this basic architecture and some of these affect the design of knowledge bases.

RECORD FORMAT

Data may be stored on a CAFS disk in the form of files of variable length records. Each record consists of a collection of self-identifying CAFS fields. The CAFS 800, which became a product, marked the end of each record with a defined field. This record terminating field is called a trailer or trailer field. There can be several trailer fields defined and they may be referred to as Trailer1, Trailer2 etc.

The CAFS field is of a varying length, and it is possible to define a CAFS field as a group of fixed length fields. This can be achieved by taking advantage of the masking facility and identifying a fixed length field as a CAFS field identifier and a mask pattern. The CAFS field identifier is called a field group identifier which is common to a set of mask patterns. Each pattern is related to a fixed length field. The last field in a field group can in practice be variable length. The data and trailer fields (see Figure 4.10) contain first an identifier code, which represents a field group identifier, followed by a field group length. In the original experimental CAFS (MK1) the beginning and end of each record was marked with field groups with reserved identifier codes. The later development of trailer fields only to mark records has several advantages, one of which is improved throughput, another is simplicity and a third is the possibility of introducing data 'punctuation'. Data punctuation is described on p.105.

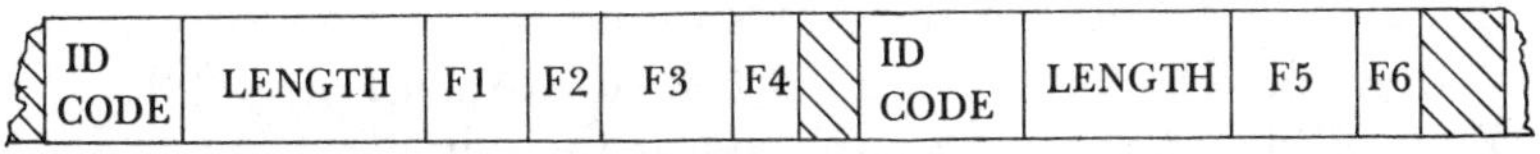

ID CODE	LENGTH	F1	F2	F3	F4		ID CODE	LENGTH	F5	F6	

Figure 4.10 *Schematic representations of CAFS fields on disk*

The disk controller contains special key-matching hardware which is loaded with a set of field group identifiers (ID CODE), the masks in the form of field offsets (start of

F1 etc) and field lengths, as well as the values of each fixed length field (values of F1, F2 etc). The key registers are thus more complex than suggested in Figure 4.9(a). A microprogram to evaluate a record selection expression of one or more keys (the pattern elements) is loaded into the search evaluation unit. This program is entered when the end of record is detected by the hardware, and if the result is True for that record, then (and only then) are the selected fields (eg see Figure 4.10 F5 and F6) returned to the mainframe. This process can be represented by the statement:

GET F5, F6, *FOR* (F1 < 3 *OR* F2 > 100)
AND (F3 = 36 *OR* F4 *PRESENT*)

JNF DATA STRUCTURES

As can be seen from this description of the CAFS records, there is no concept of levels or structure other than File, Record, Field Group and Fields. The CAFS is concerned with 'flat' files which have the primitive structure of a table. The question arises as to what are the best methods of storing and retrieving information using a CAFS for the data whose external form has a structure more complex than a table; a form that would be best placed in a network. Extended Relational Analysis (ERA) has demonstrated that highly structured data can be 'normalized' to a set of 'flat' files (tables) and the extra 'structure' given as an Sfd graph embedded as a set of procedures in the data management system. It is obviously possible to store each table as a CAFS file where selection and projection are automatically performed by the hardware via the process of selection and retrieval. However, the join of two or more tables would have to be done within the mainframe [in a version of the CAFS (Babb, 1979), special joining hardware is available (see p. 108)].

One technique for overcoming this problem is to perform all the possible joins (except self-join because of the multitude of possibilities) of the tables for an application into a single table. This process, called Joined Normal Form (JNF) (Babb, 1982), results in a table that is the join of all the tables through common attributes along the Sfd implication arcs. The n-tuples (rows) that are lost owing to the absence of common values in shared attributes can be restored by joining them to dummy n-tuples with undefined values in the non-specified attributes. The result is then appended to the single joined table. This is a physical realization of the universal relation with 'blanks'.

Each of these n-tuples of the new compound table is formed into a single CAFS record. The attributes and their associated values are stored in each record as fields. In order to recover the original tables by using the CAFS selection hardware, an extra field (called the header) at the beginning of each record contains a bit vector. The bits (or tags) in the vector indicate the tables to which the record can provide a new row of information. This vector is called the table-membership vector. The procedure for creating each vector is to scan through the single JNF table and set a bit to 1 in the vector position representing a particular projection (Boyce/Codd Normal Form or BCNF table) whenever a new combination of values is present. The vector is used to provide non-redundant answers to enquiries without any expensive mainframe processing. Each bit of the vector is a field, and this field is included in the selector. Those records that have only some of the attributes and represent information that would have been lost if only the full join had been stored can now be selected from their appropriate tables. To illustrate the construction of JNF consider, for example, the three relations P, S and PS.

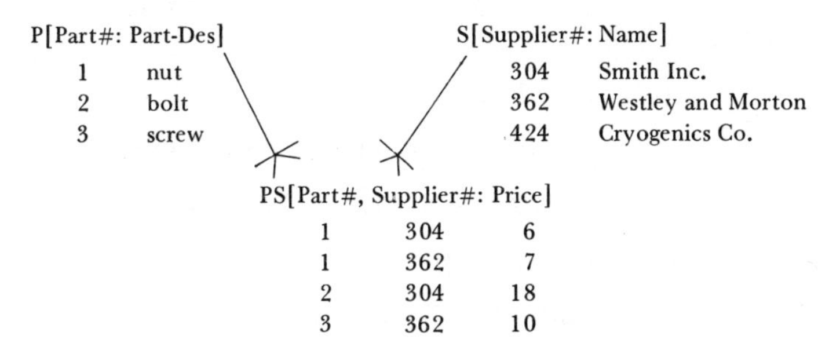

P[Part#:	Part-Des]
1	nut
2	bolt
3	screw

S[Supplier#:	Name]
304	Smith Inc.
362	Westley and Morton
424	Cryogenics Co.

PS[Part#,	Supplier#:	Price]
1	304	6
1	362	7
2	304	18
3	362	10

These would be stored as the table W, which is the join of all the tables Q and T where:

$$R = P \ \mathit{JOIN} \ S \ \mathit{JOIN} \ PS$$

and:

$$Q = P - \Pi_P R$$

which are the parts not linked to suppliers and in this case there are none. Also:

$$T = S - \Pi_S R$$

which are the suppliers not linked to parts, for example, 424. Then:

W = R *UNION* Q *UNION* T

and the extra vector field header would appear as the set of records shown in Table 4.1.

header			attributes				
P	S	PS	Part#	Part-Des	Supplier#	Name	Price
1	1	1	1	nut	304	Smith Inc.	6
0	1	1	1	nut	362	Westley and Morton	7
1	0	1	2	bolt	304	Smith Inc.	18
1	0	1	3	screw	362	Westley and Morton	10
0	1	0	–	–	424	Cryogenics Co.	–

Table 4.1 *JNF of P, S and PS tables*

Thus, the CAFS operation

GET Part#, Part-Des *FOR* Header-P = 1

will return the Part relation as originally defined and:

GET Supplier#, Name *FOR* Header-S = 1 *AND* Name = CRYO←

will return (where ← masks end of value):

Supplier#	*Name*
424	Cryogenics Co.

IMPROVING SEARCH

Differencing

One of the obvious drawbacks of such a system is the considerable expansion of data that can occur due to the repetition of values. The expansion can be greatly reduced by a process called 'differencing', where if a series of records (JNF-tuples in this case) happens to have a field with the same value in each record, then this field is not repeated in subsequent records that immediately follow the first in the series. The physical layout of the CAFS file then becomes equivalent to the normal hierarchical record structure which has records with repeated groups. The records are still present but of shorter length. This differencing reduced the physical size of the databases examined by a factor of about 2.5. (Differencing is sometimes called 'remembered hits' because the hardware has to 'remember' the value of a field for each subsequent record that does not contain that field. This 'remembering' is done by locking the chosen comparator units so that they are only reset when the associated field is encountered.) The CAFS form of this structure has the further advantage that new information can be added without any reorganization, but at the cost of less efficient use of file space. The optimum ordering of the records is usually obtained by sequencing on the primary key (eg P or S), which has the minimum number of members. Secondary sequences can then be chosen with the next compound key that has the next minimum membership, and so on. Optimum sequencing is also strongly influenced by how the data will be used in the task domain.

Variable trailer

The physical layout of the field/value pairs of differenced records can be exploited to use the basic CAFS architecture to respond to certain problems involving intersection. For example:

GET Part#, Des *FOR* Supplier# = 304
AND Supplier# = 362

will from a relational point of view require, in the purchasing database, (see Figure 4.3) the following computationally expensive processing:

((PARTS *JOIN* CATALOGUES) *WHERE* Supplier# = 304)
[Part# Des]
INTERSECTION((PARTS *JOIN* CATALOGUES) *WHERE*
Supplier = 362)[Part# Des]

The differenced CAFS 800 file for the elements involved could be:

Head/Vect, Part#/1, Des/Nut, Sup/304, Name/GKN, Price/6, Trailer1
Head/Vect   Sup/362, Name/ABC, Price/7, Trailer1, Trailer2
Head/Vect, Part#/2, Des/Bolt, Sup/304, Name/GKN, Price/18, Trailer1, Trailer2

In the CAFS 800 the records are marked only by the Trailer. However, it is possible to redefine the trailer to be one of ten (say) different record terminators that can be associated with each task. In this example Trailer1 would represent the normal termination of a JNF record. Trailer2 on the other hand could be chosen to be the record terminator for another task, and in this case the 'first record' would contain both Sup/304 and Sup/362. It is then possible to set up a task to scan Trailer2 records and retrieve the required Part# in a single operation without the need to intersect.

The nesting of TrailerNs would be done according to rules similar to those of differencing. Unfortunately, it is biased towards answering certain questions for the same reasons as the hierarchical storage structures (see p. 87). Thus, the query:

GET Supplier, Name *FOR* Part# = 1
AND Part# = 2

cannot be answered in one task with the above choice of ordering. This must be resolved by intersection in the central processor unit.

Search area indexing

It has been assumed, so far, that the disk is exhaustively scanned by the heads during a search task. However if an ordering can be placed upon the physical layout of the data that relates to the data's content, then there is no reason why advantage should not be taken of this ordering to reduce the volume of data examined. The emphasis on physical layout is more concerned with '*not* searching where the information definitely is *not*'. Thus, the effective area of disk scanned during search can be greatly reduced by exploring only those areas where the information required is likely to be. The system that controls the scanning of the disk must record the pattern of data layout, and this record is encapsulated in two kinds of indexing.

The ordering that gives optimum compression of data by differencing can be further exploited for both sequential and inverted indexing. As in Banerjee *et al.* (1978) and Lefkovitz

(1969), the advantage of CAFS indexing is that the index only refers to a search area and not to individual records. The search area, called a sucket (search bucket), contains many records (of the order of 100 to 1000). If the records are sorted, then an index of the range of values is kept for each of the 500 (say) search areas (in the experimental system there were only 200). Thus, when a selector contains this field, only the search area that contains its value (or search areas if more than one value) is scanned by the CAFS. The size of the sucket and the ordering of records can be made optimum for a particular task domain through an assessment of the data and the data's use.

Inverted indexing, on the other hand, employs a vector associated with a fixed range of values. These values are masked (or hashed) so that they represent a range of values, and each range of values of these unordered fields is associated with a 500-bit vector: one bit for each search area. If the bit is set (ie equals 1), then a record containing that field and value may be found in that search area. The vectors are retrieved from a small CAFS file and processed according to the logic of the selector. This processing is the simple disjunct and conjunct of the bit vectors according to the selector. The result is a vector which shows the series of potential search areas (resultant suckets) that are worth scanning. Figure 4.11 illustrates an example were the request is:

LIST Part# *FOR* Price = 150 *AND* Sup# = 120

The CAFS file is in JNF and has eight search areas. The file has been ordered on Part#, and in this case the numbers are evenly distributed throughout the range. Normally the sub-ranges of each search area are different since the file is sequenced and then examined for indexing. Sup# and Price have inverted indexing, and this relies upon the data having structure and hence bias. Consequently, Sup# and Price tend to have their ranges clustered. The vector (where the most significant bit is on the right) indicates the areas to be scanned for each sub-range of Sup# and Price. The two vectors that are related to the sub-ranges that include Sub# = 120 and Price = 150 are conjoined (*AND*) and the resultant vector is a vector of search areas to be scanned for this request (see Figure 4.11).

JNF CAFS file			
sucket number	PART#	SUP#	PRICE
0	0 to 100		
(1)	100 to 200	*	*
2	200 to 300		
3	300 to 400		
4	400 to 500	*	
(5)	500 to 600	*	*
6	600 to 700		*
(7)	700 to 800	*	*

inverted indexes

SUP#	sucket vector
0-100	00101100
100-200	01001101
200-300	10100010

PRICE	sucket vector
0-100	10110000
100-200	01000111

REQUEST

LIST PART#
FOR PRICE = 150 (01000111)
AND SUP# = 120 (01001101)

resultant suckets = (01000101)
ie *sucket numbers = 1, 5 and 7*
The vectors have the least significant bit on the left

where * shows the sucket indicated by a search area vector

Figure 4.11 *A simple example of the use of indexing with CAFS*

Hardware joins
It became practical and economic to consider the use of solid state memory in bulk. Certain attributes, such as Part#, Supplier#, Colour, Type have a domain that is a defined range of discrete values. It is possible to devise a hashing algorithm for each attribute so that the range of values will map into the address space of a solid state memory. This memory need only be one-bit wide (each address can store only 0 or 1) if it is to be used for set intersection; intersection can be used to implement the natural join.

The memory used for CAFS intersection was called BITMAPS (256K 1-bit wide stores) and the complete system was called the File Correlation Unit (FCU). The FCU uses the bit maps to link two (or more) files together through a common attribute value. The two files can be the same so that self-joins may be done. For example, in the case where the two files are ORDERS and SUPPLIERS then the request:

GET Supplier#, Name *FOR* Orderdate = 3/7/84
AND Address = London

and where the tables are:

ORDERS [Order#: Orderdate, Supplier#]
SUPPLIERS [Supplier#: Name, Address]

then the CAFS tasks would be:

1. *SET* Supplier# *FOR* Orderdate = 3/7/84
IN ORDERS

which sets the bit map by inserting a 1 in each address hashed by the retrieved supplier#s. The next task uses these bits as part of the selection criteria.

2. *GET* Supplier#, Name *FOR* Address = London
ANDSET Supplier# = 1
IN SUPPLIERS

Thus, all those suppliers that are in London and have a bit set in the bit map due to task 1 are retrieved. *ANDSET* is used here to indicate a reference to the bit map. *IN* indicates the file to be searched.

Similarly, the problem involving intersection of the same file (or self-join) can be resolved:

1. *SET* Supplier# *FOR* Part# = 1
IN CATALOGUES

2. *GET* Supplier#, Name *FOR* Part# = 2,
ANDSET Supplier# = 1
IN CATALOGUES

The advantage of hardware joins using the FCU is that the hashing into a bit map sorts the linking attribute 'on the fly', and the result can then be used to restrict the search of the next task without any referral to the central processor unit. Further, the data may be stored compactly and with minimum repetition (except for the linking attribute values) in a manner consistent with relations. This produces a uniform method of answering queries without any bias. It can be shown that if the bit map has a smaller range of potential values than the range of values for the linking attribute, then the result is extra records that do not conform to the required selection criteria. However, no records of the required type are lost. In practice, these extra records represent a small percentage of the total. The expected error is 50% of the possible error for a single attribute join. This error reduces rapidly the more join attributes are involved so that after four joins the expected error is less than 10% of the possible error.

Commercial and practical pressures on the CAFS have forced the need to operate on standard files and databases. This is possible because a field group (see p. 101) can have the image of a normal record generated by a COBOL program, for example. The self-identifying format is only used for text. Further, it has been found, in practice, under the conditions of information retrieval from standard databases, that multihead reading is not needed, and this search power is better employed for parallel accessing of disks. CAFS can now be used for both content and direct addressing (mixed mode).

The choice

The choice of database design depends upon both the available machinery and the tasks to be done. Considerable efficiency and speed can be achieved through traditional techniques that employ direct addressing. Difficulties arise when changes occur in the task domain, and these changes then influence both the data structure and the material to be accessed.

The advantage of the conceptual model is that it distinguishes between the task domain and machine specifications. Some of these choices can be made automatically. It is possible for conceptual databases of a restricted kind to be stored in both an address mode and a content accessing mode so that the most efficient access method can be used. The current (1984) performance in the mixed mode is a 30 to 90 times improvement in elapsed time and a reduction in central processor unit usage (number of instructions) by about 16,000 times (ICL CAFS Presentation Notes, 1984). This approach is worth considering for on-line real-time use for very large knowledge-based (or database) systems. However, for small knowledge-based systems that are complex and stable, many of these large-scale problems are irrelevant since most of the system can be kept in main store.

The stages of design have involved a sequence of models where each model requires new decisions to be made. In summary the models are:

1. the conceptual model (the derivation of which requires several stages;
2. the logical model;
3. the physical model.

The models help the designer make judgements by restricting the issues to those of the same type without distorting the design objectives. The conceptual model may seem to be concerned only with the task domain, but the method of representation has been influenced by the logical model. The logical model, in turn, reflects the possibilities given in the physical model. However, there is a characteristic of the conceptual model that also captures the client's language within the task domain. Chapter 5 illustrates how this characteristic may be used to improve man-machine dialogue.

CHAPTER 5

A Conceptual Database Interpreter

> . . . he that hath words of any language without distinct *ideas* in his mind, to which he applies them, does, so far as he uses them in discourse, only make a noise without any sense or signification. . . . For all such words, however put into discourse according to the right construction of grammatical rules or the harmony of well-turned periods, do yet amount to nothing but bare sounds, and nothing else.
>
> *John Locke (1689)*
> *An Essay Concerning Human Understanding III.X.26*

Introduction

This chapter illustrates a simple interpreter called the Flexible Language Interpreter (FLIN) which supports an interactive Flexible Interrogation and Declaration Language (FIDL). FIDL is a semi-formal English language used by a user to interrogate and modify a database. The interpreter FLIN embodies a conceptual model of a database in order to construct 'explicit' Content Addressable File Store (CAFS) tasks from 'implicit' user commands. The commands are implicit because they specify only the user's objectives and not the details of the steps necessary to satisfy these objectives. The implementation details are described elsewhere (Addis, 1973, 1975b, 1976, 1982a).

The man-machine dialogue depends upon a primitive definition of understanding such that a message is acceptable (understood) if a translation can be found that is consistent with the conceptual model. This is effective because the conceptual model embodies the language and grammatical structure of the task domain. If a representation is coherent with respect to the way in which a group of people talk about the tasks, then every statement that has at least one reading which is interpretable with respect to this model, should also have meaning for the group.

System outline

Figure 5.1 shows the outline of the FLIN system. The FLIN Data Definition Language (see Appendix 1) allows the conceptual model to be specified for the task domain. This specification is divided into two parts. The first part defines the Semantic functional dependency (Sfd) graph, and the second part defines the method of storage and access. These specifications are compiled into an internal representation (which is stored on the disk with the associated files), and it is read into main store at the time of calling a particular application via the interrogation program.

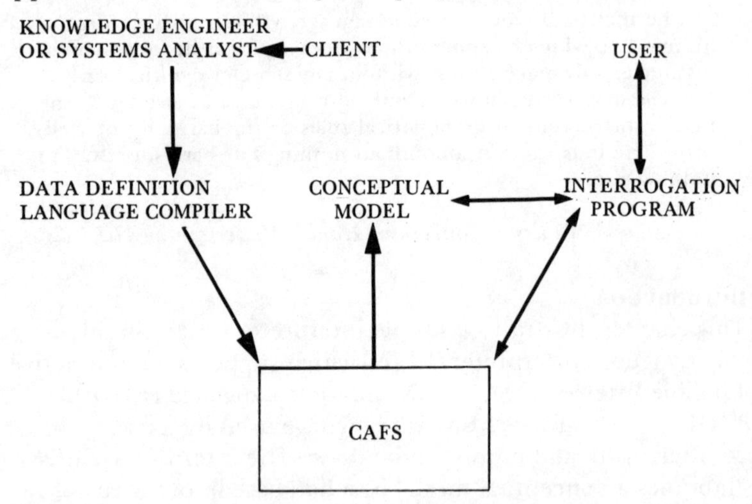

Figure 5.1 *Outline of the FLIN system*

FLIN is the interactive interpreter for FIDL. FLIN uses a conceptual model to translate user's sentences into a set of CAFS commands. The interactive language FIDL (see Appendix 2) allows the user to ask for a list of data items or to count certain data items which are related by some logical combination of predicates. An example of this is:

LIST ALL ORDERS *FOR WHICH* PART *IS* SCREW *OR* BOLT

The user does not need to know anything about the relationship between the data items or how the information is stored. Admittedly, the language FIDL is not relationally complete. FIDL can only deal with predefined relations; certain kinds of joins are not possible (cf Sharman, 1975;

Codd, 1971a, b, c, d). Consequently, many legitimate queries are not allowed. However, those queries which are permitted can be written more simply than in a full relational language (see Chapter 4).

The language also allows new information to be put into the database. The interpreter FLIN uses a conceptual model to validate all new information and to construct questions to put to the user for any missing information not already given or found in the database. Only information which conforms to the conceptual model can be put into the database. Conversely, any information removed from the database is also validated so that subsequent enquiries by a user always produce consistent results.

Updates are initiated by the function words *CREATE*, *CHANGE* or *DELETE*. The consequences of the update are known from the Sfd graph. Information which cannot be retrieved from the database is requested from the user by the program. All the checks described in Chapter 3 are performed during the interaction. For a simple example (cf Figures 3.5 and 4.3):

CREATE ORDER

ORDER-NO ? = *8362*	Checks ORDERFORMS for given ORDER# This order number is not found
SUP-NO ? = *6231*	Checks SUPPLIERS for given Sup# which is found

ORDER-NO 8362
INSERTED

The mechanism of implicit queries

In the Jobs Database (see Figure 5.2) the query '*LIST ALL* Part-No, PD, Town *FOR* Sup-No 19876' can be interpreted by taking the attribute names used (in this example Part-No, PD, Town and Sup-No); and by working backwards along the directed graph from the tables in which these attributes are members, the first common table found on the traced paths is the PS table. The PS table is called the 'root' table. [Note that this particular conceptual model is different from that given in Chapter 3 (see Figure 3.8), since Figure 5.2 does not involve join dependency.] A root table is a table which implies all other tables which have to be *joined* in order to answer (project out) a particular query. The root to which the user is referring can be deduced from the attribute

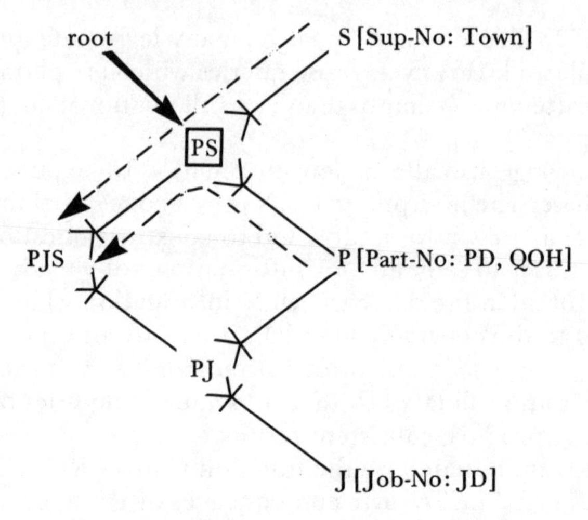

Figure 5.2 *Find the 'root' table for S and P*

names and the structure of the query. The root table PS shown in Figure 5.2 would generate the table:

PS *JOIN* P *JOIN* S

From this table the requested information would be projected:

(PS *JOIN* P *JOIN* S) [Part-No, PD, Town]

If the tables are stored in Joined Normal Form (JNF) with membership vectors, then the joins are already precompiled. If CAFS is used, then the membership vector can be used to select on the root table and the retrieval registers will perform the correct projections.

The reason for the root rule for estimating the tables (needed to answer a query by the user) is that the normalized tables are assumed to represent entity sets and the functional dependencies represent connections between these entity sets. For the list of entities in a user's query, the interpreter should assume the most immediate connection between them – and this corresponds to the root rule (cf Preference Semantics in Chapter 10, p. 282).

In practice, the interpretation of requests becomes more complicated in order to account for logical connectives. These connectives are found in the *selector*. The selector is that part of the sentence which specifies by means of predicates the constraints to be imposed on the information retrieved. For example:

... *FOR* Sup-No = 19876

or:

... *FOR* Sup-No = 19876 *OR* Job-No = 231

A *retriever*, on the other hand, is that part of the sentence which specifies the attribute values to be returned to the end-user. For example:

LIST Part-No, Sup-No ...

The rule for determining the root tables is to conjoin the retriever and selector and then transform the result into a disjunctive form. Each resulting conjunction is considered as an independent set of command parameters. The simple root rule is then applied to each conjunction separately. For example:

LIST Part-No, Sup-No *FOR* Sup-No = 19876 *OR* Job-No = 231

will form two possible roots:

(Part-No and Sup-No) and (Sup-No) the PS root table
(Part-No and Sup-No) and (Job-No) the PJS root table

This is interpreted as a compound question since it refers to the two root tables PS and PJS. Another example of use of the root rule is shown in Figure 5.3.

TABLE	REPRESENTS
P [Part-No: Description]	The parts supplied
S [Sup-No: Name, Address]	The supplier of parts
PS [Part-No, Sup-No:]	The catalogues of the supplier
O [Order: Sup-No]	Orders to suppliers (Order forms)
PO [Part-No, Order: Item-No]	Parts on order
OD [Order: Date]	Deliveries (Advice Notes)
POD [Part-No, Order, Date: Qty, Price]	Order lines indicating the details of each order

Figure 5.3(a) *The specification of the purchasing system*

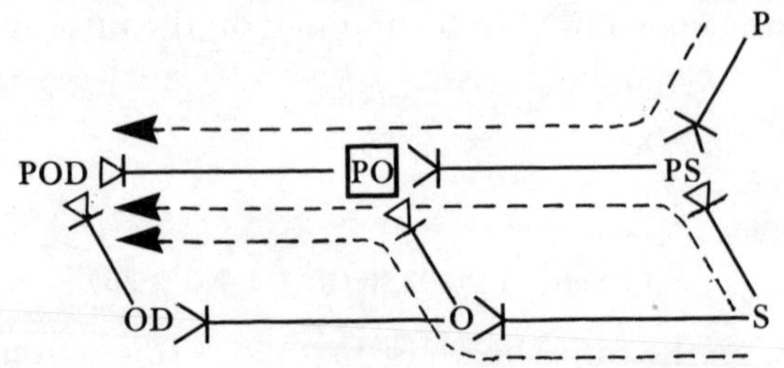

Figure 5.3(b) *The Sfd graph of the purchasing system*

In this example the request is:

'*LIST ALL* Orders *FOR* Supplier Jones *AND* Part-No 3622'

and the root tables are deduced by considering each attribute name in the query and listing each table that implies it. These are:

Order	←	O, PO, OD, POD
Supplier	←	S, PS, O, PO, OD, POD
Part-No	←	P, PS, PO, POD

The interpreter applies a function to select the root table that effectively traces the dotted lines in the Sfd graph (see Figure 5.3b). Another function then chooses the most-implied common table, which in this case is PO.

If the selector had been 'Supplier Jones *OR* Part-No 3622', then this would generate two root tables – PO and O – and there would then be two answers (two tasks).

Each set of roots is found by applying a function to the disjuncts of the selector and the retriever. The union of these results is then taken. In general, the selector is resolved into disjunctive normal form, and each disjunct is treated in turn with the retriever. The FLIN program implemented rejected queries which referred to more than one root.

The advantage of FIDL is that the user expresses his request in familiar terms and FLIN interprets it in the context of the conceptual model which embodies the user's linguistic competence of the task domain.

Cardinality constraints in FLIN are represented separately, and they are illustrated here by special graphs. The links (mappings) between table name and key or own attributes are also needed, and these are shown by other graphs (see Figures 5.4 and 5.6).

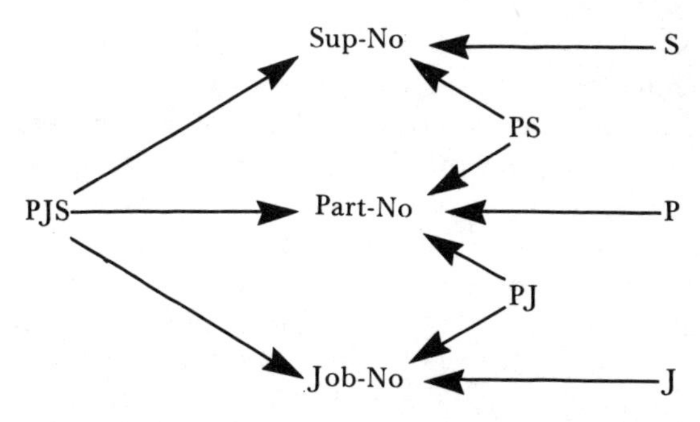

Figure 5.4 *The links between tables and their keys*

Mechanics of insertion

To insert new data the user types a list of attribute names after the word *CREATE*. This list should be just sufficient to identify the entity desired for inclusion. The interpreter then determines automatically the root table(s) in the same way as for the retriever in the *LIST* command (see p. 115). All tables implied by the root must also be updated. Figure 5.5 will be used to illustrate this *CREATE* process. Note that other kinds of constraints can be separately specified, and these influence the process of updating.

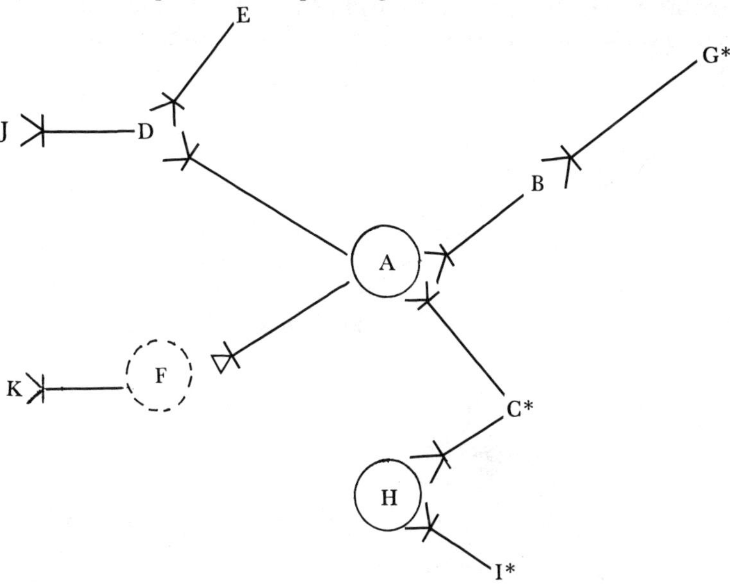

Figure 5.5 *An Sfd graph used to illustrate CREATE*

The asterisk (*) in Figure 5.5 shows the tables whose 'own' attributes have been chosen to identify the update entities. These entities are deduced to be (A) and (H). However, the table(s) which 1-implies the root(s) must also be updated (F). The total set of tables to be checked includes all those implied transitively by F, which are F, A, B, G, H, C, I.

Next, it is necessary to check that the tables concerned contain the correct data by generating CAFS tasks much as is done during interrogation. If a table does not contain data appropriate to the update, then this data must be acquired from the user and inserted before the update can be allowed to proceed.

The complete process at the task level takes the form of a dialogue. The data required is supplied either by the database or the user; it is then inserted by adding it to the CAFS records under construction. The records will not be inserted into the CAFS file until complete.

The user is required to provide only that data which cannot be deduced. To ensure that the user information is kept to a minimum two further mappings need to be defined. These are the connections between tables and their KEYS and the links between tables and their OWNS. It is the key attributes which are *not* also own attributes in other tables that are the ones whose values are usually required from the user.

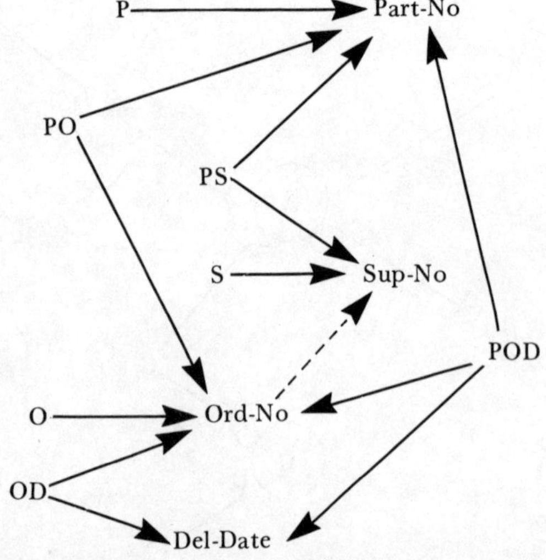

Figure 5.6 *The links between tables and key attributes (K) for the purchasing system*

Figure 5.6 illustrates the links between tables and key attributes for the purchasing system. The dashed arrow indicates the link that is determined through comparison with the links between the tables and their own attributes. Thus, if a tuple in PO is to be updated, then the keys Part-No and Ord-No would initially be required. The Sup-No can be retrieved via Ord-No from O, and if it is found it can then be employed with Part-No to check PS and so on. This procedure minimizes the set of attribute values requested from the user. The attributes are printed one at a time to the user who responds by giving the associated values. These values are then checked to ensure that they are of the correct type.

The tables to be searched, in order to retrieve the keys for other tasks, are those which contain OWN attributes that are also KEYS (foreign keys). A CAFS search task is made of these tables using the keys given by the user thus retrieving the KEYS that are OWNS. If these attributes are not found in the database, they are then requested from the user together with the OWN attributes for those entities to be inserted. The other set of tables to be checked is that set of tables whose key attributes are not own attributes, and if the values are not found in the database the values are then requested from the user. For efficiency, tables are checked by setting up a CAFS task which is satisfied by the first hit. The order of doing the tasks is from the root and checking in the direction of the n-implies (0- or 1-implies). If the task is satisfied, then any further searches in the direction of the n-implies are also assumed to be satisfied (by definition). Note that the information is stored on the CAFS in the FLIN system as a JNF table with a table-membership vector.

Hence, every record contains values of all the implied attributes, and thus the own attributes are retrieved from the first hit. The result of an update task is that the system generates a single CAFS record containing all of the attribute values collected from the database and the user.

Examples of insertion

The behaviour of the system can be further illustrated by the following three examples:

1. To create a new row (entity) in a table (which is bound to other tables by 0-implications only) requires a search for rows (entities) in these other tables implied by this new row. Consider, for

example, the situation where a user wishes to assign a salesman to an established customer in the ORDERS application (see Figure 5.9). The user can then type either:

CREATE CUSTOMER SALESMAN

or if he is familiar with the table name he can type:

*CREATE CUS**SAL*

The user's responses are set in italics and square brackets [ ] denote comments about the process.
CREATE CUSTOMER SALESMAN [The user types in his requirements]
ODB5 MAKE CUSTOMER SALESMAN [The system responds by specifying the particular file accessed]
THE ASSUMED TABLE IS CUS*SAL (CUSTOMER SALESMAN)
[The root is deduced]
CUSTOMER = *123456* [and the essential key attribute values requested]
SALESMAN = *654321*
CHECKING CUS*SAL WHERE CUSTOMER = '123456'
SALESMAN = '654321'
NONE FOUND [The particular tuple to be inserted does not currently exist and since it contains no own attributes none are requested]
CHECKING SAL WHERE SALESMAN = '654321'. OK [The salesman is already recorded]
CHECKING CUS WHERE CUSTOMER = '123456'. OK [The customer is already recorded]
OK. 1 CREATED [The completed record containing all implied information is added to the CAFS file]

2. When 1-implies is involved during creation the entity is treated as though it belongs to the tables which 1-implies the table being updated. There is the peculiarity that FLIN will still check the table to which it belongs, even though there is already evidence that it must exist as in the following example:

CREATE CUS EQP
ODB5 MAKE CUS EQP
THE ASSUMED TABLE IS CUS*EQP (CUSTOMER, EQUIP)
ORDER = *12345678* [Despite the root being CUS*EQP the key attributes for the 1-implying ORD*ITEM are asked for]
ITEM = *1*
CHECKING ORD*ITEM WHERE ORDER = '12345678'
ITEM = '1'. OK [The 1-implying row (entity) exists]
CHECKING CUS*EQP WHERE CUSTOMER = '123456'
EQUIP = '9123456' [However, the check on the implied row (entity) is still done]
ENTITY ALREADY EXISTS
NONE CREATED [Because it exists no record is created]

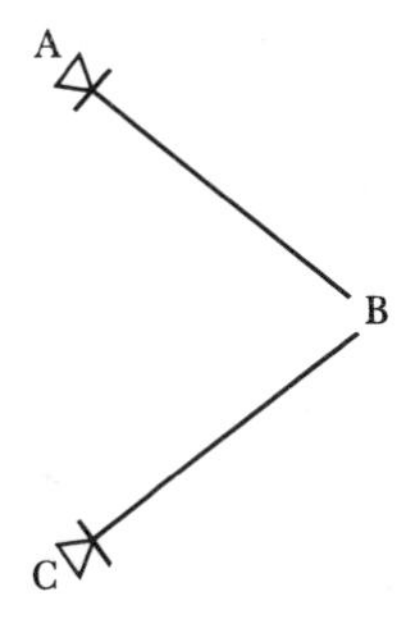

Figure 5.7 *Converging 1-implications*

In the case where there are 1-implications converging onto the table to be updated such as B shown in Figure 5.7, the program FLIN will treat the data as belonging first to A and then to C with the final check of B. The decision as to the creation of the record depends entirely on the existence of the entity in B.

3. Having collected the key attribute values for the table to be updated, the technique employed for transitive constraints is first to deduce all the key and own attributes implied by this table. If there are any own attributes which are also key attributes, then somewhere in the Sfd graph there is at least one transitive constraint. The next step is to determine the tables whose own attributes contain keys of other tables. With the information already obtained from the user, these own attributes which are also keys are collected from the database ready for the validation process. If they cannot be found, the information is obtained from the end-user. The order in which these tables are checked is in the reverse direction to that of the implication arrows. Once all these tables have been cleared, the process continues as before but with suitably reduced checking. In this way, the example given below of *CREATE* would add 'mathematics' to the list of possible 'subjects' taught in the 'course' geography. The following describes this process by the creation of the first row (entity) of STUDSUB in the student database.

TABLE	REPRESENTS
COURSE [cname:]	The different courses provided by the college
SUBJECT [sub:room, lect . . .]	The different subjects taught
STUDENT [sen:name, cname. . .]	The students that attend the college. In this case a student can only belong to one course at a time
SUBCOR [sub, cname:]	The subjects taught on each course
STUDSUB [sen, sub:]	The subjects taught to students

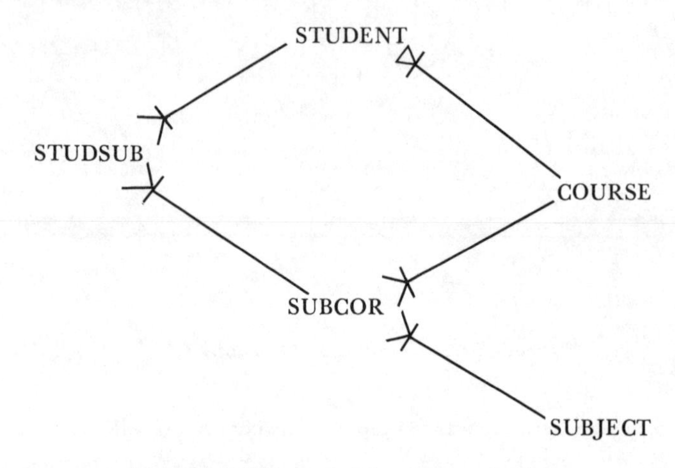

Figure 5.8 *Sfd graph of a student database*

CREATE SEN SUB
ABCD MAKE STUDSUB [The file is called ABCD]
THE ASSUMED ENTITY SET IS STUDSUB (SUB, SEN) [The deduced root relation]
SUB = *MATH* [The collection of keys]
SEN = *2*
CHECKING STUDENT WHERE SEN = '00002' [Checking STUDENT first because it contains the required key 'cname']
NONE FOUND [Because it is not there the own domains are requested]
NAME = *FRED*
CYEAR = *1*
TUTOR = *BARLOW*
CNAME = *GEOG* [In this case cname is collected from the user but if SEN = 2 had existed this information would have to be obtained from the data base]
CHECKING STUDSUB WHERE SUB = 'MATH' SEN = '00002'
NONE FOUND [The root is now checked and not found]
CHECKING SUBCOR WHERE CNAME = 'GEOG' SUB = 'MATH'
NONE FOUND [The check on this can now be done with the full key because of the previous task to SN]
CHECKING SUBJECT WHERE SUB = 'MATH'
NONE FOUND [Own attributes of SUBJECT requested]
ROOM = *12*
LECT = *SMITH*
TIME = *1203*
DAY = *MON*
CHECKING COURSE WHERE CNAME = 'GEOG'
NONE FOUND [No COURSE 'GEOG' so one will be created]
OK. 1 CREATED [The record is created]

Examples of deletion

To delete from a table requires that deletes must occur where necessary from all other tables which imply that table. The interactive process using techniques similar to insertion is illustrated by the following two examples:

1. The selector (and the retriever if there is one) interact to specify the table whose row is to be removed. Because the deletion is done on physical records these remain on the file, and delete simply removes (sets to 0) the appropriate bit in the table-membership vector. In the following example some rows in the EQP table of the DEMOB application (see also Appendix 1) are required to be removed (see Figure 5.9).

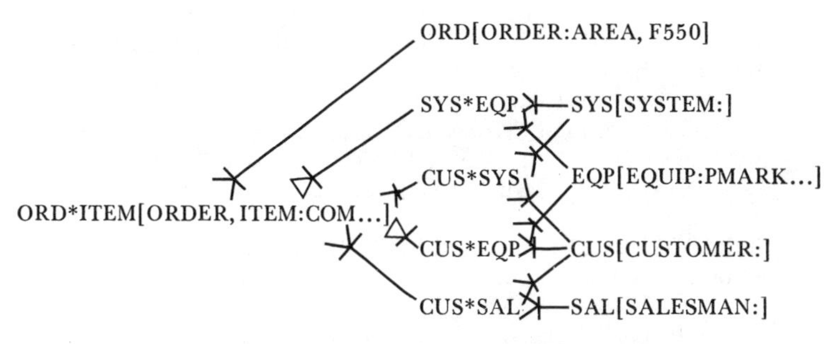

Figure 5.9 *Sfd graph for the Demob Database showing some attributes*

DELETE ALL FOR EQUIP 123450 OR 123457 OR '123456'
[ALL or some limit of records to be deleted must be given as a precaution. The assumed limit is 1]
ODB5 DELETE ALL FOR EQUIP EQUALS '123450' OR EQUIP EQUALS '123457' OR EQUIP EQUALS '123456'
THE ASSUMED TABLE IS EQP (EQUIP, PMARK, MACHINE, BAR) [The EQP table is deduced]
THE DEPENDENT TABLES ARE CUS*EQP SYS*EQP ORD*ITEM [The tables affected are given]
2 FOUND. DO YOU WISH TO CONTINUE. *PLEASE* OK [The number of records found to be deleted is given, and as a further precaution some affirmative reply is required]
OK. 2 DELETED [The appropriate bits are set to 0 in the membership vector]
The expanded interpretation is given which calculates the table EQP for deletion simply from the domain names used. After this the ORD*ITEM, CUS*EQP, SYS*EQP implying EQP tables are listed to show that rows in these tables, if associated with the rows to be removed, will also be deleted. Note that the word 'dependent' means *either* the implied tables (for insertion) *or* the implying tables (for deletion). Only two physical records are found and after the checking question these two records are deleted.

2. A more complex example is now considered in which the 1-implies constraint to CUS*EQP and SYS*EQP from ORD*ITEM in the DEMOB application (see Figure 5.9 and Appendix 1) is checked and kept. In the previous example the 1-implies was automatically maintained because all three tables could be traced backwards on the same paths specified by EQP. In general, the key attributes to the 1-implied entities have to be collected from the CAFS records before these records are deleted. Consider a case where a salesman is deleted who has sold two 2903 machines to a customer who has two systems. Now if this salesman is deleted from the SAL table, then the implying rows (entities) in CUS*SAL table must also be deleted and so also must the implying rows (entities) in ORD*ITEM table. However, if these deleted ORD*ITEM rows (entities) are the last ones associated with the systems and equipments in the SYS*EQP and CUS*EQP tables, then these rows must also be deleted in order to maintain the 1-implies constraint between them and ORD*ITEM (see Figure 5.9). The program would operate as follows:
   DELETE ALL FOR SALESMAN 543261
   ODB5 DELETE ALL FOR SALESMAN EQUALS '543261'
   THE ASSUMED TABLE IS SAL (SALESMAN) [The deduced table is SAL]
   THE DEPENDENT TABLES ARE CUS*SAL ORD*ITEM [The implying tables are found]
      COLLECTING KEYS FOR CUS*EQP AND SYS*EQP [The CAFS record that specifies the salesman being deleted also contains the values of the key attributes of the 1-implied tables CUS*EQP and SYS*EQP. The values are collected and stored]
   2 FOUND. OK [Only 2 CAFS records are found that contain different information]
   3 FOUND. DO YOU WISH TO CONTINUE. *YES* OK [However, 3 records are required to be deleted in total]
   OK. 3 DELETED [The 3 records have the SAL, CUS*SAL and ORD*ITEM bits set to 0 in the membership vector]
      CHECKING ORD*ITEM FOR CUS*EQP WHERE CUSTOMER = '123456' EQUIP = '0290300'
   NONE FOUND [Two of the key attributes collected previously that identify a CUS*EQP entity are used to scan the records associated with the ORD*ITEM table in order to see if there are any ORD items implying this entity. None are found so a delete task is set up]
      DELETING CUS*EQP [Therefore delete forward along 1-implies]
   THE ASSUMED TABLE IS CUS*EQP (CUSTOMER, EQUIP)
   THE DEPENDENT TABLE IS ORD*ITEM
   OK. 1 DELETED [The CUS*EQP relation bit is set to 0 in the membership vector]
      CHECKING ORD*ITEM FOR SYS*EQP WHERE SYSTEM = '212122' EQUIP = '0290300'

NONE FOUND [Again, another task is set up to check to see if the ORD*ITEM row deleted was the last one associated with the key attributes collected. It is the last one]
DELETING SYS*EQP [Deleting occurs forward along the 1-implication as before]
THE ASSUMED TABLE IS SYS*EQP (SYSTEM EQUIP)
THE DEPENDENT TABLE IS ORD*ITEM
OK. 1 DELETED [The SYS*EQP bit is set to 0 in the membership vector in the record answering to the above description]
CHECKING ORD*ITEM FOR SYS*EQP WHERE SYSTEM = '654321' EQUIP = '0290300'
NONE FOUND [Yet another SYS*EQP row is checked as before. Note that this completes the 2 records deleted in the first instance]
THE ASSUMED TABLE IS SYS*EQP (SYSTEM EQUIP)
THE DEPENDENT TABLE IS ORD*ITEM
OK. 1 DELETED [The last SYS*EQP bit is set to 0 in the membership vector in this other record]

In the last example, three CAFS records were associated with the salesman 543261 (Personnel Number). Only two of these CAFS records contained key attribute values to the 1-implied tables CUS*EQP and SYS*EQP. It can be concluded that the third CAFS record is a lone record [not linked with associated entities and requiring only the SAL row (entity) of the salesman to be deleted]. The other two CAFS records have the ORD*ITEM and CUS*SAL bits set in the table-membership vector but only after the key attributes in these CAFS records of the 1-implies tables have been collected. These key attribute values are employed in the unusual CAFS task of scanning the CAFS records with the selection on the ORD*ITEM table (membership vector bit set) and the key attribute values of the 1-implies table CUS*EQP or SYS*EQP. It is unusual because it does not obey the root rule. This task is repeated three times with different values, and each time no records satisfied the selector. The ORD*ITEM rows deleted must have been the last ones implying these rows (entities) in CUS*EQP and SYS*EQP, so they in turn are deleted by removing the appropriate bits.

In comparison with the insertion of rows, the 1-implies always has to be checked in the reverse direction to that normally done for the 0-implies. The meaning of this deletion task is that the salesman has *never* existed. This is why his sales vanish when he does. To prevent this kind of nonsense, the SAL table must only be allowed to be added to and never be deleted from. The program provides this facility by checking all affected tables in an update against a 'No additions' and 'No deletions' vector associated with the system. If this constraint is about to be breached, then the update is not allowed.

Interaction of language and the conceptual model
Since the interpretation of requests both for query and update depends upon the internal representation of the conceptual model then the meaning of the commands will be affected by this model. A good example of this is the effect implication constraints have on the meaning of the functions *MAKE* and *DELETE*. If all the implication constraints consist of 0-implies, then *MAKE* and *DELETE* are not always the inverse of each other. This is because the insertion of a row entails the insertion of other rows in tables that are not affected by the deletion of that same row at a later stage. This means that the database is not left in the same state as before the act. However, it was found that most users assumed the actions to be the inverse of each other, and this could lead to some unexpected results.

One possible solution is to provide a *RESET-BEFORE* command which would re-establish the previous situation. Apart from requiring a considerable overhead in recording the history of the database so that it can be re-established, this is still not what is always meant by *DELETE* after *MAKE*. Another interpretation is to assume a function *ERASE* which behaves like *DELETE* except that every implication is taken as a 1-implies. In this way, every trace of the existence of a row can also be erased.

The effectiveness of the conceptual model
The FLIN system was an experimental tool which has been tested by on-line users for several applications with databases ranging from 4 MCh to 30 MCh, and it was found to be both versatile and robust. The response of the system to a complex query on an ICL 1903A using the CAFS MK1 under the GEORGE III operating system with a normal work load ranged from 1 to 20 seconds for these applications. The simplicity of the Data Description Language (DDL) has enabled an analyst using ERA to design and modify a conceptual model with considerable ease. FLIN can take the resulting model

and provide an immediate enquiry and update language which is flexible enough to satisfy a broad spectrum of users from clerks to managers.

In considering the needs of the user, who may wish not only to interrogate the database but also to add new information, it was necessary to extend the conceptual model to include other constraints. In particular, the cardinality constraints have been particularly important in the updating of a database, and can be considered similar to the domain constraints on attributes in that each can be stored or computed by algorithm. These constraints enable a user to interact with a better model of the perceived world which is automatically maintained by a machine. The advantages are that the user does not have to specify his requirements in unfamiliar terms, and that logical inconsistencies are never permitted. Furthermore, interpreters of the kind described here (which have a parametric description of a database) eliminate the expensive need for a multiplicity of programs to fulfil the same requirements. The fact that the model is parametric means that different users can see the same database in many alternative ways thus automatically providing a comprehensive privacy system.

Users found that the automatic interpretation was particularly useful, and in some cases the users were surprised at the answers given by the system. The surprise was caused mainly by those users not fully conversant with the task domain, and, in this sense, the program was behaving as a teacher. It would have been easier for the occasional user if each attribute could have been specified by a descriptive noun phrase, rather than by a unique name. This was because such users could not always remember the attribute name but only its role. The man-computer conversations lacked a certain 'naturalness' caused by the fact that each FIDL statement was an isolated formula to be interpreted. However, despite these shortcomings, the ease with which the general user learnt to operate the system, and remembered how to communicate with it even after a long absence, showed that a full natural language is not always necessary or appropriate (Chapanis, 1975; Kelly and Chapanis, 1977).

Limitations of ERA

Extended Relational Analysis (ERA), as given, is a representation scheme that has been created in order to express the important components of a task domain. In particular, it provides a means of describing entities by assuming that they form well-established categories (entity sets) such that each member of a category has a common behaviour pattern. This behaviour is defined in terms of constraints. The resulting analysis embodies the sub-language and grammar of the task domain.

The components that form the analysis are semantic functional dependencies, relational algebra and set operations. The semantic functional dependencies describe constraints between the entities in different entity sets, and do not express functional dependencies between attributes. In general, the entities in each set may be complex, structured and have no attribute in common. However, the elicitation procedure (see p. 68) assumes that the Sfds can be inferred from shared attribute names. This assumption does not always hold. For example, the attribute name 'Part#' may refer to parts in stock, parts in the catalogues, supplier's part numbers or parts used on jobs. The precise use of an attribute must be questioned by the analyst before the semantic functional dependencies are recorded.

The relational algebra and set operations (such as *PROJECTION* and *JOIN*) require some knowledge of the attribute domains. Similarly, the set operations (eg *PARTITION*, *UNION*, *INTERSECTION*) need certain limitations on the structure of the entity sets. Several problems arise because of these limitations, and many of the problems are due to the restrictions imposed by normalization. Some of these problems are: key domains with overlapping dependencies that cannot be normalized satisfactorily (Sharman, 1976), the unnatural formation of some entity sets needed to retain atomic entities, and the containment within the normalization procedures of functional interrelationships between attributes in the same and different tables (eg in [a, b, c] where a + b = c, and total). However, many of these problems are containable within the ERA development programme.

Work began on ERA in 1972. ERA is a hybrid method involving two analytic principles. The first principle concerns entity sets and Sfd whereas the second principle concerns domains (as for attributes) and set theory (eg join, union and partition). ERA (an amalgam of the two principles) is incomplete (in the light of recent research) and is context-dependent in its interpretation. It is, however, of great practical use because it reflects the language of the client, is easily sketched during consultation and has proved its utility during elicitation at the initial stages of contact. Further, ERA, through the creation of a conceptual model, has a central role in the design of large scale knowledge-based systems.

Current research work (1984-85) suggests that the two analytic principles should be used as two distinct phases of design. The first phase (called ERA1) should involve only entity sets and Sfd. This first phase would be used for sketching out a conceptual model during the primary consultations with the client. The second phase (called ERA2) would be extended to include many more set operations (eg projection, union, intersection, difference, etc.) and thus allow precision analysis. Associated with these set operations are transformations which describe how a database changes in response to updates. ERA2 identifies more exactly the activities and components of a knowledge-based system (see Grant and Elleby 1984 for an illustration of ERA2).

This projected two-phased approach is not described since it is still under development. For the purpose of future distinctions, the hybrid method given in this book is called ERA1+. Chapter 7 illustrates how ERA1+ can be used to define a program that solves specified problems within a task domain.

The conceptual model provides three clear aspects which are not made distinct in ERA1+. The first aspect is TRUTH TESTING where the *extension* of the model is compared with the current facts, and the response is TRUE or FALSE. The state of affairs, as given by the tables, is either correct or incorrect. The second aspect is TRUTH MAINTAINING where the *intension* of the model is used to transform one extension of the model to another. This transformational role is used to track the changes in the state of affairs or explore possible states derivable from a given situation. The third aspect is TRUTH SELECTION where the intension of a query is matched with the intension of the model to construct the extension of the query from the extension of the model. The FLIN system performs these last two aspects.

ERA1+ does not capture very much of the problem-solving component of a task domain. However, most problem-solving acitivities within a task domain can be expressed as relational operations on the extension of the conceptual model. Other techniques such as Petri Nets (Peterson, 1981) have also been used to describe the flow of activities within an organization. The conceptual model interacts with this flow of activities to give a competence model of the 'experts' in the task domain. The next chapter illustrates a competence model of diagnostics determined using ERA1+.

CHAPTER 6

A Conceptual Database Consultant System

> 'Now,' said Rabbit, 'this is a Search, and I've Organised it –'
> 'Done what to it?' said Pooh.
> 'Organised it. Which means – well, it's what you do to a Search, when you don't all look in the same place at once.'
>
> *A. A. Milne (1928)*
> *The House at Pooh Corner*
> *Methuen*

Introduction

Consultant systems are usually knowledge-based systems incorporating a single mode of reasoning. Nevertheless, information retrieval systems of the kind discussed in Chapter 5 are consultant systems of a kind. The major problem with retrieval systems is that there is very little guidance offered as to what questions need to be asked in order to solve a problem. ERA1+ says little about problem solving, since it is concerned mainly with representing the fabric of the perceived world with respect to a task domain. It captures one aspect of the users' competence but ignores the users' performance of the tasks. Figure 6.1 illustrates the extra dimension required to construct a knowledge-based system that incorporates these users' skills. The result is a competence model that explicitly represents the entities and the rules through which the problems are solved.

Medical diagnostics has been described as three distinct processes (Young, 1979) acting in two stages. The first stage consists of stepping through a fixed sequence of tests (or questions) looking for a pattern of symptoms that matches some known diagnosis. This is tantamount to a database search. However, the initial pattern does not usually lead directly to a final diagnosis but will cause further patterns to be inferred and matched. If this pattern matching process

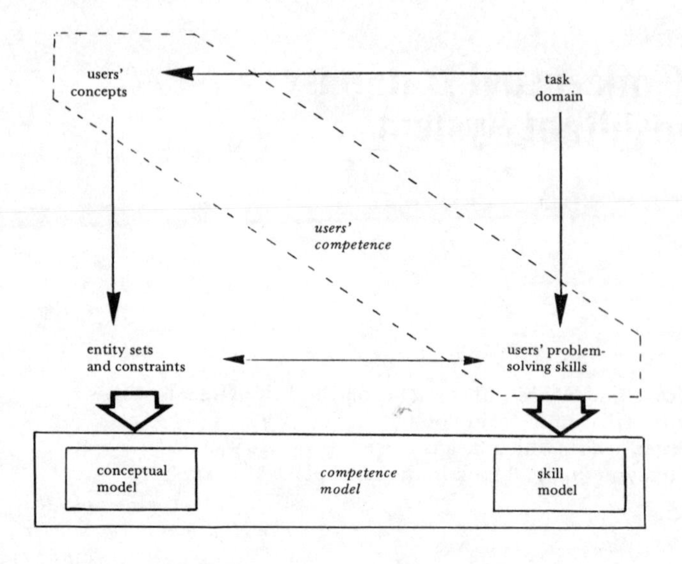

Figure 6.1 *The competence model (cf Figure 3.2)*

fails, then this stage continues with the fixed sequence of tests. If no diagnosis is achieved by the end of the test sequence, then a problem-solving process is entered that relies upon hypothesis and inference. This new process also searches for known patterns.

The conceptual database can be used for simulating the first stage. In the early 1970s Brunel University, working in conjunction with ICL, proposed a diagnostic computer fault-finding technique based upon pattern matching of symptoms within a relational database (Hartley, 1984). After some initial trials, the system was later reformulated in terms of a conceptual database and implemented on the CAFS MK1 (Addis and Hartley, 1979; Addis, 1980). Since it was known that at least 75% of hardware faults and 65% of system software faults have previously been encountered by engineers in the field, much effort could be saved by providing a system that could guide an engineer quickly to a similar fault. However, a straight information retrieval system does not help. There needs to be an approach to retrieving information that is linked closely to the diagnostic process such that the optimum search is related to fault observation and investigation.

The CRIB system

The Computer Retrieval Incidence Bank (CRIB) system was initially conceived as a diagnostic aid for the engineer, by providing him with an accumulation of symptom patterns gleaned from many sources, including his peers. The important aspect of this collection of symptom patterns is that it is used to generate further questions, according to certain rules, so that the engineer can be quickly guided to the fault indicated by the symptoms. It is this feedback, above all, which differentiates this kind of diagnostic aid from a typical information retrieval system. Thus, the engineer gives the system a description of his observations, and these descriptions are compared with the database. Groups of symptoms which match the incoming description are processed to produce a list of the best possible tests for the engineer to make next.

The CRIB system was devised as a means of accumulating information on critical symptoms used for diagnosis. This idea of accumulating fault reports for computers is not new, and it was used in the early 1960s for the ATLAS computer at Harwell. What was new in the CRIB system was both the use of the conceptual model to maintain a database of diagnostic reports and the extensive use of 'action' assessment procedures to help the engineer find the fault.

The CRIB system consisted of two programs. The first program was the conceptual database interpreter FLIN (Flexible Language Interpreter) which used the CAFS (Content Addressable File Store) as its main means of data storage and access. The second program retrieved data from the CAFS, and assessed the next best tests to be done in order to home-in rapidly on the current fault. Thus, instead of the user (engineer) having to generate queries, the system presents the user with recommendations as to the kind of query that will retrieve the fault that fits the observations. To do this, the system requests the user to provide information on the results of applying significant tests. What is a significant test depends upon the contents of the database and the context.

A fault is considered to be both a replaceable unit and a set of symptoms, where a symptom is a test with a result. A test can be as simple as making an observation. These tests are referred to as actions. Each test is predefined and indexed by a number prefixed by the letter A. The central feature of the CRIB system is that the action is designed to elicit symptoms from the patient machine via the engineer. Thus, the action code A1036 means 'run store test program X', and this results in the symptom 'store works' or its inverse 'store faulty'. The symptoms can be expressed either in code (eg S1036 or the inverse N1036) or in jargon English as above.

The action-symptom pairs (symptoms for short) are organized in groups which can be of three types:

1. total group: the accumulated group of symptoms observed to occur during many investigations of a single fault (T-group);
2. key group: a subset of symptoms of a total group all of which are necessary to indicate the location of a fault (K-group);
3. subgroup: a subset of symptoms of a key group which, by virtue of being an incomplete key group, can only indicate the fault location to a degree less than certain.

The contents of symptom groups reflect the experience of actual investigations, and a group only exists if it has proved its usefulness in actual fault investigations. In particular, there is a floating population of subgroups which are compared for usefulness by a simple success ratio (the fraction of times the symptoms in the group have occurred in successful investigations relative to the total number of investigations in which the subgroup occurred). The more successful subgroups are considered as active, while the less successful ones are kept in reserve.

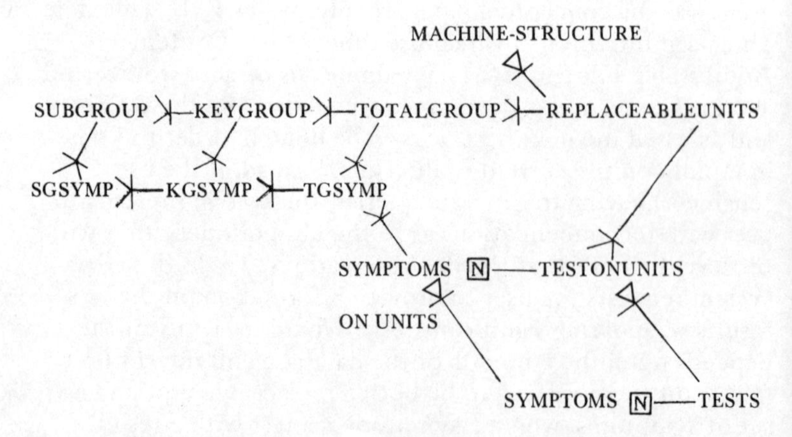

$$\mathrm{CARD}(A, \boxed{N}, B) \Rightarrow (\exists f \in F_S)(\forall b \in B)\{(A \supseteq \overleftarrow{f}(b)) \wedge (2 = |\overleftarrow{f}(b)|)\}$$

Figure 6.2 *Sfd graph for the CRIB diagnostic aid*

Figure 6.2 illustrates the conceptual model for this diagnostic procedure. The final Extended Relational Analysis (ERA) shows that there are only a few entity sets involved. A test is just an independent atomic entity that is associated with only two results (SYMPTOMS). There is no means of determining the similarity of tests or expressing any operational equivalence. The interpretation of these tasks is left to the user.

When a group of symptoms occur together (ie as observed by the engineer) during a sequence of actions or tests, then this group is associated with that part of the hardware in which the fault lies. Each replaceable unit is constructed out of other replaceable units. The machine is modelled as a simple hierarchy of subunits (as in a parts explosion) and expressed as a simple binary relation (MACHINE-STRUCTURE). An example of the first few levels is given in Figure 6.3.

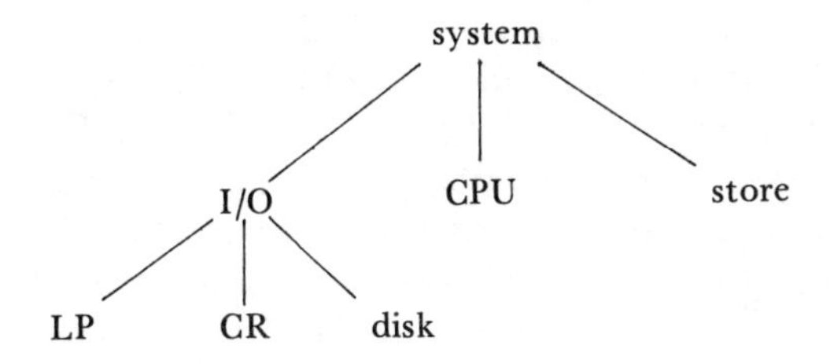

MACHINE-STRUCTURE [Unit, SubUnit:]

system	I/O
system	CPU
system	store
I/O	LP
I/O	CR
I/O	disk

Figure 6.3 *A simple hierarchy of subunits of the patient computer*

The purpose of the model is not to represent a 'true' picture of the organization of the machine, but to provide some structure to the database so that a search can be ordered. The searching of this tree is depth-first where a decision to drop a level is only made on the basis of matching a complete group (total, key or sub in that order) from among the set of symptoms observed by the engineer. With only 50 to 100 nodes on the tree (subunits in the hierarchy), this is manageable.

By successively matching more symptoms, the CRIB

system will eventually arrive at a terminal subunit which will be either repairable or replaceable. With the trend towards increased modularity, most fault investigations are eventually an exercise in board swapping. If a terminal subunit is reached and the fault is not cured, this can mean one of four things:

1. a path taken based upon a chosen subgroup did not lead to the fault;
2. a key group is incomplete (more symptoms are necessary to be certain of success);
3. multiple, transitory or 'soft' faults – these are difficulties in any diagnostic system, including manual ones;
4. the hierarchy is incapable of reflecting the fault location sufficiently precisely.

The system then backtracks automatically to the last decision point, and tries to find another match. Since the search is depth-first, the whole hierarchy can be covered if necessary, ie every symptom group in the database will eventually be examined. If this happens and the fault is still present, then it is a fault which the system has not seen before. When the system fails to find a match with even a subgroup at any level of the hierarchy, it asks for fresh information rather than backtracking immediately. This is not done in a blind fashion, but via a list of suggested actions put to the engineer, chosen on a heuristic basis using information about the diagnostic context current at that time. Several heuristics have been tried, but all are aimed at attempting to complete incomplete subgroups, thus enabling a drop in level, ie a more precise location of the fault. From the partially matched subgroups at that level, symptoms are compared for their usefulness according to four criteria:

1. the time taken for the associated action to be carried out;
2. the average amount of time the investigation is likely to take after this step has been taken;
3. the number of incomplete subgroups in which this symptom occurs;
4. whether the symptom is the last one in a subgroup.

These factors are combined in a heuristic procedure to order the relevant actions which have been extracted. The five best actions are suggested to the engineer. The engineer is not obliged to carry out any or all of these actions, but having carried out one or more actions and having reported the

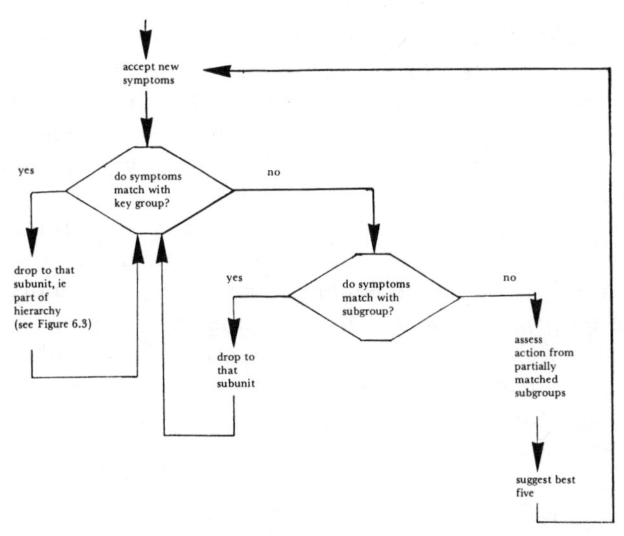

Figure 6.4 *Flow diagram of the CRIB diagnostic process*

symptoms observed, the search cycle is then repeated. Figure 6.4 shows the diagnostic process without the backtracking.

An alternative view of this classification process is proposed by Johnson and Keravnou (1985). Their view draws upon Peirce's three stages of inquiry and the terminology of Pauker *et al.* (1976). Puzzling phenomena may be used to infer by ABDUCTION a certain hypothesis (ie a hypothesis is 'generated' to account for the phenomena). Through DEDUCTION certain experimental consequences of this hypothesis can be arrived at. On carrying out these experiments, some of which may not tally with the deduced results, a new or modified (proposed) hypothesis is formed through INDUCTION (ie the hypothesis is 'generated' *before* the tests). Potential hypotheses in the expert system Present Illness Program (PIP) (Pauker *et al.*, 1976) are classified as INACTIVE, ACTIVE or SEMI-ACTIVE.

Using these ideas, a status transition diagram (see Figure 6.5) can be constructed for the CRIB system where hypotheses are statements about faulty units. Initially all hypotheses would be inactive but puzzling phenomena will be observed. The most general hypothesis would be activated through the process of 'abduction' – SYSTEM FAULTY (see Figure 6.2). This hypothesis suggests through deduction (via the MACHINE-STRUCTURE) that one of the subunits is faulty, and also suggests a set of potential symptoms. These hypotheses are assessed, and the one worth pursuing is made active while the others remain semi-active. Hypotheses rejected are made inactive. This view provides a terminology for the process implied by the Lakatos Programme of scientific theory formation (see p. 69). An abduced hypothesis must have the progressive component such that the deducible consequences are derivable from experiments and the deducible consequences must include also the puzzling phenomena (hence the hypothesis 'explains' the observations). Induction can only be justified if the experiments are considered fairly chosen and the hypothesis to be tested is decided beforehand.

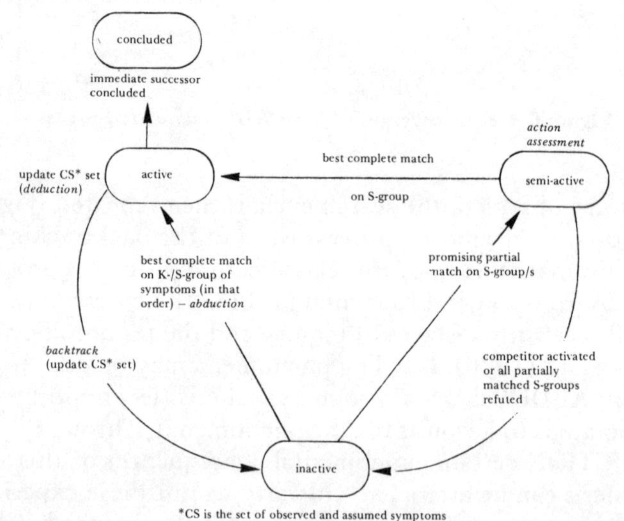

Figure 6.5 *Hypothesis status transition diagram (from Johnson and Keravnou, 1985)*

CRIB on CAFS

DESCRIPTION

The initial CRIB system was hampered by the lack of an efficient data retrieval system. It was for this reason that the experimental CAFS MK1 was used, which led to some interesting changes in principle, implementation and performance.

The early system operated as a depth-first search procedure where a match was sought only at the level below the current subunit (hypothesis), among its associated groups. Thus, for all the symptom groups in the database, only those relating to the immediate subunits were examined. In this way the required search was contained.

CAFS, however, is capable of searching large databases globally, without any extra effort. Using this facility, the CRIB system can now look for a match with all symptom groups at all levels. The notion of a depth-first search then disappears. It is now possible for a fault to be located within any appropriate subunit in one step, instead of using the original step-by-step, level-by-level procedure. Since this is the case, the notion of backtracking out of unhelpful situations becomes unnecessary. Because of this major change in thinking, the organization of the program was altered. In particular, group matching and action retrieval techniques were revised.

In the original CRIB system there were four files (created through combinations of tables), organized according to the required major search tasks. Speed of access was limited by the need to scan these files repeatedly. On the other hand, the CAFS MK1 system depends upon all the data existing in one large file, where each CAFS record contains one example of each kind of data item (a Joined Normal Form or JNF file). This means that each symptom of a particular symptom group is contained in a separate record, and all the symptoms of this group are scattered throughout the file. Each record also contains a table-membership vector, which indicates its membership of one or more of the tables. Each table also represents the natural join of each of their implied tables, thus providing a considerable amount of precompiled processing of tables. Eight of these tables are shown below. They are directly drawn from the conceptual model (see Figure 6.2).

TESTS [A: TTP, NA, LOT, DES] — is the list of all possible tests and actions an engineer can perform. This contains information on how long each test or action takes (TTP), number of times used (NA) and level of training required to perform the task (LOT). DES is a character string describing the test or action.

SYMPTOMS [S:A]	forms a list of all the symptoms and observations made by engineers in performing a particular test.
TOTALGROUP [T:SU, CTS]	this represents the sets of groups of symptoms associated with each subunit. They are called total groups. CTS is an attribute showing the total number of symptoms associated with each total group.
TGSYMP [T, S:]	this shows which symptoms belong to which total groups.
KEYGROUP [K:T, CKS]	is the set of necessary and sufficient symptoms within the total groups that can isolate a replaceable unit. These are called key groups. CKS is an attribute showing the total number of symptoms associated with each key group.
KGSYMP [K, S:]	this shows which symptoms belong to which key groups.
SUBGROUP [G:K, SUCC, NG, SIT, CGS]	is the set of necessary but not sufficient symptoms within the key groups. They are called subgroups. This also indicates the success of a particular subgroup in tracing the correct replaceable unit (SUCC), number of times used (NG) and investigation time (SIT). CGS is an attribute showing the total number of symptoms associated with each subgroup.
SGSYMP [G, S:]	this shows which symptoms belong to which subgroups.

The CAFS MK1 hardware cannot, in itself, discover whether there exists a key or subgroup which is a subgroup of the observed symptom set (more than one symptom). However, this can be achieved by both a software simulation of bit maps and an extension of the principle of the bit maps called count maps. For the purposes of effective use of the maps, the target identifier values (in this case the group identifier) must be transformed via a hashing algorithm into a map address (each bit or count word in the map corresponds to one value).

All codes in the CRIB system, such as symptom and group identification, are four-digit numerals which can be used as relative addressing to a word in store. This store is equivalent to the bit map in the CAFS file correlation unit except that each element is eight bits instead of one. Each hit of a particular address is 'added' to that element. This store is called the count map, and it is used to count the number of symptoms in a group. Let KI and GI refer to the four-digit key group and subgroup numeric identifiers, respectively. CKS, CGS are the group counts, ie the number of symptoms in the groups. Sx refers to a member of the observed symptoms (ie one of the symptoms observed by the engineer translated into the four-digit code).

The CAFS tasks:

(1) *LIST* KI, CKS *FOR* (S=Sx) *IN* KGSYMP

(2) *LIST* GI, CGS *FOR* (S=Sx) *IN* SGSYMP

are performed for each of the observed symptoms Sx. The retrieval group identifiers KI and GI each point to a word in their respective count maps, and add in one to the count word associated with each group found (see Figure 6.6).

The numbers in the count maps, after all the symptoms have been dealt with, represent the number of symptoms observed by the engineer that are found in each group, and for comparison the maximum number of symptoms possible in each group (CKS and CGS) are also retrieved (the fact that in some cases it is retrieved several times does not matter). A match is found if the number in the count map is greater than zero.

If a match is found, the remaining details of subunit (and key group in the case of subgroup match) are needed. There are two more tasks:

(3) *LIST* SU *FOR* (KI=Kx *OR* Ky *OR* . . .) *IN* KEYGROUP

(4) *LIST* SU *FOR* (GI=Gx *OR* Gy *OR* . . .) *IN* SUBGROUP

Here Kx, Ky . . . and Gx, Gy . . . refer to those groups which have been successfully matched. It is conceivable that more than one subunit will be found, although such ambiguities are not desirable since the whole idea of a key group is to identify uniquely a particular subunit.

KI → K1038 (+1)

	key group count	count map CKS
K1036	0	–
K1037	1	5
K1038	4	5
K1039	2	6
K1040	0	–
K1041	3	7

GI → G1038 (+1)

	subgroup count	count map CGS
G1036	0	–
G1037	1	3
G1038	3	4
G1039	2	6
G1040	0	–
G1041	1	3

Figure 6.6 *Examples of count maps*

To minimize the search, those subgroups which are partially matched but which contain at least one symptom which is the inverse of an observed symptom, can be eliminated from further consideration. The task to do this is:

(5) *LIST* G *FOR* (S=N1 *OR* N2 . . .) *IN* SGSYMP

where N1, N2 etc are the inverse symptoms of the observed symptom set.

The action details for unmatched symptoms in partially matched subgroups are required for assessment. The count map used in (2) gives the partially matched subgroups (see Figure 6.6), ie those with a count greater than zero and not equal to CGS. For each of these groups Gx, a task is needed:

(6) *LIST* A, TTP, DES
FOR G = Gx *AND* [S#≠(S1 *OR* S2 *OR* . . .)] *IN* SGSYMP

where S1, S2 . . . is again the set of observed symptoms.

AN EXAMPLE OF THE USE OF THE CRIB SYSTEM

Given that the CRIB system is ready for use, the engineer will identify himself and then report the initial observed symptoms. This can be done by a set of cryptic English statements or by an internal code. To illustrate the protocol, consider an actual example of an engineer using the system to find a fault. The user's responses are italicized and comments are given in square brackets.

LOG IN FRED, HERE
INVESTIGATION STARTED AT 10.35 ON 12.11.78
PROGRAM LOAD FROM DISK NOT OK
SYMPTOM ACCEPTED AS N0022. ANY MORE SYMPTOMS?

PROGRAM LOADED FROM CR NOT OK
SYMPTOM ACCEPTED AS N0026. ANY MORE SYMPTOMS?

ALTER AND DISPLAY ONES OK
SYMPTOM ACCEPTED AS S0402. ANY MORE SYMPTOMS?

NO [Having accepted three symptoms, the program starts its search and makes the following responses]

SUBGROUP MATCH FOUND IN CPU
[Action assessment starts]
42 SUBGROUPS EXAMINED
9 SUBGROUPS ELIMINATED
59 ACTIONS ASSESSED

SUGGESTED ACTIONS [The active hypotheses] :
* CHECK FOR A PARITY ERROR – A0008
* LOAD MPD WITH 2 LOG MOVES – A0415
* CHECK HAND SWITCHES INDIRECT – A0418
* RUN WAMS – A0389
* CHECK RUN LIGHT – A0413

[Here five actions have been selected as being the most effective ones from the 59 which were considered (the semi-active); MPD and WAMS etc are codes which the engineer would understand]

SO N0008, N0415, S0413 [This illustrates the short form of reporting observed symptoms where the 'A' of the action code given is replaced by S or N depending upon the result of the test]

ANY MORE SYMPTOMS?

NO [The second cycle begins]

SUBGROUP MATCH FOUND IN CPU

68 SUBGROUPS EXAMINED [Note that there is still insufficient information to allow the CRIB system to isolate the fault any further although the search has been widened]

39 SUBGROUPS ELIMINATED
58 ACTIONS ASSESSED
SUGGESTED ACTIONS:
* CHANGE BOARD 9713-A0405
* CHECK HAND SWITCHES INDIRECT-A0418
* CHANGE BOARD 9716-A0412
* CHANGE BOARD 9710-A0422
* CHECK STEPS OF DISK IPL ROM-A0406
LOG OUT OK

In this case the first suggested action (A0405) cleared the fault. The engineer was satisfied the diagnosis was complete and terminated the run as OK]

INVESTIGATION FINISHED AT 10.40 ON 12.11.78
DURATION OF INVESTIGATION 4.5 MIN

ASSESSMENT OF THE CRIB SYSTEM

There were several difficulties which arose out of the use of the CRIB system. The first problem was perhaps the tendency of the CRIB system to spread its considerations over more and more groups every time a symptom was added. This had the double effect of, first, slowing the response because of the greater number of tasks to be generated, and, second, the insertion of apparently irrelevant actions into the suggested list because of it making contact with distant groups. This problem really reflects the method used in the CAFS search, and could have been contained by using the hierarchical model (see Figure 6.3) to constrain the search to the relevant part of the machine under consideration.

A second problem was the uncertainty of how to answer some actions to select the desired symptom. This problem falls into three parts:

1. The jargon English interpreter, although perfectly adequate when dealing with a handful of contextually dependent sentences, was found to be incapable of handling a large range of open-ended descriptions. The engineer could never be sure of how to phrase a sentence so that it would be accepted, or if it was accepted whether it had been identified correctly. However, users of the system quickly learnt a few starting sentences, and then employed the returned action code with the appropriate prefix to continue.
2. Some answers to an action were ambiguous. If a symptom was observed there could be uncertainty as to whether a positive or negative response to the suggested action was expected.
3. This is really a consequence of the second part. There was no way in which a set of symptoms in a new group (indicating a freshly discovered fault) could be checked to see if they already existed. Trial and error, scanning down all the existing symptoms or using some kind of selector mechanism based upon the concatenation of words in a sentence (keywords) might help, but they are all uncertain methods of checking. There was also the related update problem in that no guarantee could be given that there existed a distinguishing symptom in each group that would ensure isolation of a fault. This last problem could be overcome by employing the diagnostic techniques during update.

Despite these difficulties (most of which have solutions), the system proved to be successful, and, in particular, the suggested actions helped the engineer to consider areas of investigation which he would otherwise have overlooked. This had the useful result that even if the fault being investigated was not to be found within the CRIB system, it could still help the engineer isolate this new fault by suggesting avenues of enquiry.

A simplified CRIB system

DESCRIPTION

Shortly after the CRIB (CAFS MK1) system was implemented, there was the suggestion that it could be used for system software investigations. In particular, it could be used for VME/K and DME known faults by acting as the initial pre-scan before technical specialists were involved. With a considerable amount of help from those technical specialists and after several variants of the CRIB system and its database had been tested, a simplified version of the CRIB system was constructed.

Subgroups were abandoned, leaving just total and key groups. Each action was given a priority ranking instead of 'time to perform' (TTP) and marked either P (for patch) or Q (for question). Each key group (now called trace) was given a 'likelihood' which acted as a priority modifier for all the actions within the key group. This priority was used as one of the factors in ordering the suggested list of actions returned to the user. The user then had the choice of either having the returned suggested actions ordered according to the number of hits and (for those with the same number of hits) ordered according to the priority modified by 'likelihood' or having the suggested actions ordered according to the modified priority only. The user also had the choice of listing out any number of the ordered actions he wished instead of just the first five. 'Likelihood' was intended to indicate the probability of a symptom group occurring so that as the database is updated so this parameter would be adjusted for each group.

The problem of restraining the CRIB system from contacting distant groups was left in the hands of the user by providing him with two distinct selection mechanisms. The first mechanism used the count maps in three ways:

1. If the user has reported m symptoms, then he can select from the count map only those groups which have a count greater or equal to (m − n) where n is an integer of his choice. If n is zero, then this is equivalent to insisting that all the symptoms given are contained in the groups retained for consideration.
2. Irrespective of how many symptoms the user has reported, the system can scan the count map looking for the maximum number of hits and only report on those groups with this maximum number. A simple extension of this idea is to allow (Max − n) as the selection mechanism.

3. Count map selection can also be done relative to the total number of expected symptoms in a group (CKS in Figure 6.6). The selection in this case is done by subtracting n from each group number (CKS), and if the count is greater than or equal to this then it is returned. Another way of describing this method is that it selects those groups which have at least a specified number of symptoms in total. Smaller groups than this number behave in the way described first, but the smaller the group the more symptoms are ignored. The number of symptoms given by the user + n = maximum total in groups.

The second mechanism employs a modification of the unit-subunit hierarchical structure used in the CRIB system (see Figure 6.3). In this case, the hierarchical structure represented *an organized approach to diagnosis constructed from the experience of the experts*, where each level in the hierarchy represented sets of questions concerned with details of the level above. A five-digit code was constructed so that each character position represented a level, and the character chosen gave the branch at that level. Figure 6.7 shows part of this structure, where the code at each level is given in parentheses, and each level is associated with a priority number (the lower the number the higher the priority). The user is then able to confine his search to any subsection of this hierarchy by typing in, for example, SU 32B** or SU **B*D, where * represents a 'don't care' character (this is easily done with CAFS through the use of masking).

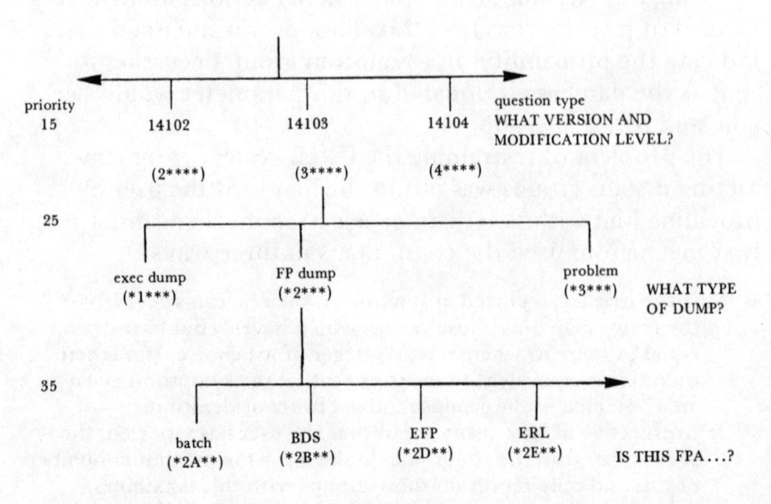

Figure 6.7 *Part of the VME/K faults database structure*

The problem of ambiguous answers to actions (questions) was solved by simply allowing only one answer to be accepted, namely, '*YES*'. Thus, the question 'is operating system version 14?' would either be answered '*YES*' or not answered at all. This decision to have only one answer leads to other problems, but these other problems are easier to cope with for this application than ambiguity.

ASSESSMENT OF THIS SIMPLIFIED CRIB SYSTEM

This simplified CRIB system is more of a tool in the hands of experts than the original CRIB system was, since many of the strategies of pattern matching are decided by the user. The speed of response was considerably improved and the probability of success greatly enhanced with most of the original problems bypassed. The only outstanding problem was that of ensuring that symptoms were not repeated in another form when updating occurred. This problem was partially overcome by classifying symptoms. Each class then consisted of a sufficiently small number of symptoms to be scanned by eye. It was intended to make an extended trial of this modified system, but conflicting demands made it necessary to curtail any further tests; however, enough was done to show the importance of involving, at quite a detailed level, the experts and the users in the development of such a tool. The final system required less than five minutes' explanation to a user for him to use it efficiently. The importance of such a tool is that it represents a medium by which an expert can express and record his expertise in a way that is self-maintaining. This is why it requires the specialist to 'create' the database.

Conclusions

The CRIB systems have the power to be truly consultative in that the user participates by choosing the next tests to be done from the proffered tests. This simple act draws upon the special knowledge that the engineer has of the context, and this special knowledge can tune the decision and pattern matching procedures used by the program. The CRIB system shows the engineer all that is possible within the limitations of the database, and the engineer adds his views through selection.

The CRIB system cannot 'explain' its actions or present a

case for a particular action because it does not have a general deductive mechanism. The transparent and simple manner in which the CRIB system works makes such requirements unnecessary. However, systems that are based upon general problem-solving mechanisms are explored in the following chapters.

CHAPTER 7

The Structure of Artificial Intelligence Programs

'Wear this (garter) for good luck.'
'He draws to the left, so lean to the right.'
'The wind is from the east, so aim to the west.'
'He crouches when he shoots, so stand on your toes.'

Advice given to Painless Peter Potter based on the film 'Paleface' (1948) (Bob Hope and Jane Russell)

Introduction

One of the major objectives of artificial intelligence is the determination of the design principles needed to create a machine that can 'think' (McCorduck, 1979). This concern has its foundation in cybernetics; a science that studies the common principles of natural and artificial systems which respond to the environment to achieve a goal. They are principles independent of physical form, and one such principle is intelligence. Intelligence is a principle that governs an important class of goal-oriented responses. Cyberneticists believed that intelligence could only be understood through the study of natural systems expressed in the mathematics of state transformations and information theory (Ashby, 1956). The mathematics of state transformations is closely related to group theory where a state of a mechanism is expressed as a set of statements (eg switch to the left, light is on) and the transformations are a set of mappings from one state to the next (eg [Left, On] maps to [Right, Off]). Experimental observations of the brain, influenced by this 'new' science were presented in terms of behavioural psychology and cybernetic theory. Models were constructed (eg Walter, 1953), unbounded by any preconceptions of computers, and restricted only by the engineering skills of the maker. These models illustrated beautifully the anthropomorphic power of simple machines that move through space according to simple and recognizable laws of animal behaviour.

Cyberneticists achieved some success in creating a general theory of machines, but they were unable to make much impression on describing the nature of intelligence in a way that could be used to engineer a system. They were able to define the necessary conditions for intelligent behaviour, but were unable to provide sufficient specifications to create an intelligent machine. They showed that the essence of goal-directed behaviour was contained in the use of negative feedback, and that the limitations of intelligent behaviour were measurable through the laws of information theory (eg The Law of Requisite Variety) – (Ashby, 1956).

Those in the field of artificial intelligence, on the other hand, abandoned the need to wait for the physiologist to unravel the circuits of the brain (such as the work of Hubel and Wiesel, 1962, on the physiology of the optic cortex, McCleary and Moore, 1965) and they decided that intelligent behaviour could best be expressed as symbol processing. This view was supported by the work of Simon and Newell in the analysis of the statements made by people when asked to give a running subjective assessment of their thinking while solving problems. It was possible to 'explain' these subjective assessments as state transformations that acted directly on the symbol strings representing the problem. What was more impressive was that in the late 1950s using primitive computers they were able to make a Logic Theory (LT) machine (Newell *et al.*, 1963) using this technique. This machine eventually evolved into the General Problem Solver (GPS) (Newell and Simon, 1963). At the heart of the GPS is the mechanism for applying transformations to symbol strings where the symbol strings represent the initial statement of the problem. The initial symbols that represent the starting conditions of the problem are transformed into different and meaningful symbol strings that give a potential route to a solution. Each transformation is linked to an acceptable step of reasoning about a problem. The purpose of the mechanism was to find a sequence of transformations that led to a 'solution' (ie a symbol string recognized to be the goal conditions of the problem).

Problem solving was taken to be the paradigm of 'thinking', and a state transformation approach based upon theorem-proving methods has been accepted as the central ideal for reasoning. There is plenty of evidence that this paradigm and ideal do not encompass much of human intellectual performance. It can be shown that people do not respond to the symbols in isolation, and they tend not to follow all the dictates of logic that are implied by the transformational approach (Wasen and Johnson-Laird, 1968). There have been some very strong arguments that suggest that artificial intelligence does not describe any aspect of human thinking (Dreyfus, 1972). This argument in outline shows that the process of insight cannot be explained satisfactorily in symbol processing terms; and since insight is the essence of 'thought', then 'thought' is not explained in these terms either.

The important issue, from the knowledge engineering point of view, is that the factual knowledge represented in the computer is complemented by some method of reasoning. The method of reasoning should be justified, accepted and expressible in a mechanistic form. It ought to make no appeal to 'understanding' but exist upon its own foundation. Further, it must be expressive enough to model successfully human performance in problem solving. The task of the knowledge engineer is to create a competence model that will give an acceptable response to a given state of affairs: a response that can be equated with expert and rational activity. It is the competence model that ultimately provides a service to people employed in a particular task domain.

Industry characterizes a set of task domains for which knowledge engineering has an important role, and for this reason the schism between artificial intelligence and industrial practice has become less apparent. This is partly because industry is demanding more complexity and intelligence from its increasingly powerful computers, and partly because artificial intelligence is reaching a stage where many of the techniques have become established. One of the reasons why these established techniques have not been adopted by those involved in industrial applications has been the radically different approach to the design of programs and systems. Industry has been concerned with large-scale problems requiring teams of programmers to be organized within a fixed managerial structure. What is required is a guaranteed product with well-specified performance, which is clearly documented for maintenance and change independently of those who initially created the product. Consequently, design methods have been an all important development within industry, with the intention of providing a uniform method of systems analysis and specification (eg Jackson, 1983).

Artificial intelligence has concentrated more upon structural complexity of a kind rarely seen in industry (Ritchie and Hanna, 1982). The programs are normally considered as experiments in complexity, and they have only an ephemeral existence. The design of these programs depends upon on-line interactive editors with easy access to interpreters for trying out parts of programs during their development. In this manner, convoluted and intricate programs are created. They are formed like works of art rather than being engineered, thus allowing individual styles and programming principles to be fully exploited. It can be argued that it is only through such interactive systems that it is possible to cope within a reasonable time with the kinds of programs required to illustrate artificial intelligence concepts.

Behind the skill, intellect and committed artistry of the programmers there is a sound collection of techniques. These techniques have been established by the 'hard' school into a methodological core from which most artificial intelligence programs are eventually shaped. This core consists of production systems, graph search, rule-based inference (based upon the Resolution Principle) and heuristic control.

Production systems

THE THREE ELEMENTS

A production system is formed from three distinct elements: the global database, transformation rules (or production rules) and control strategy. The global database represents the problem states, the transformation rules express how the database may change to conform with the problem, and the control strategy decides from all the applicable transformations which one should be applied to achieve the solution (goal state). All the rules can have access to the global database at all times, but they must not act upon themselves or the control strategy.

Each rule has a precondition that relates to the global database and describes when the rule can be used. If the precondition is satisfied, then the rule can be applied to the database to transform it into a new state. The control strategy chooses the rule by drawing upon some heuristic knowledge about the nature of the problem and its solution. It is called 'heuristic' because it is 'an aid to discovery' (ie helps find the goal). It does not necessarily have any analytic foundation but calls upon a deeper understanding of the problem than that apparent in the representation alone. The control strategy ceases computation when it reaches a termination condition; a condition where the objectives are satisfied.

The principle of the production system can be illustrated by the problem of navigating a car through a town in a foreign country whose language is unknown to the driver. The goal is to travel through the town from south to north making as few wrong turnings as possible. Unfortunately, the town is old, large and rambling with no regular structure. There is no map of the town available, but there is a compass in the car which gives an approximate indication of north.

The global database represents the current junction giving the number of turnings, the density of houses and a history of such junctions ordered according to their encounter. The transformation rules are marked by the choice of turnings at each junction. The transformation is the effect of taking a road from the current junction to the next junction. The intermediate road connecting the junctions is the mapping function from state to state. When the mapping function is (initially) unknown, it is called 'implicit'. The act of taking a route makes it 'explicit'. The heuristic for the control strategy is to make a choice that is governed by some crude generalization of this mapping. The heuristic control strategy could be, for example, 'take the turning that is nearest the northern route'. This heuristic does not guarantee a correct choice, since that route may change direction and even double back on itself. The heuristic can be further enhanced by taking into account the distribution of houses encountered at each junction on the basis that the density of houses usually increases towards the centre of town. The goal is reached when all the houses are in the south, and the density of houses at the junction is zero. There may be several such goals since there may be more than one road out of town leading north.

There are two kinds of heuristics proposed to help find a solution to these problems. The compass provides a clue to the possibility of a particular transformation being a good one to try (the pre-emptive heuristic), and the density of houses indicates the effectiveness of a transformation once it has been tried. The second kind of heuristic (density of houses) illustrates the 'generate-and-test' approach to problem solving where a possibility is explored and then matched against a requirement.

ILLUSTRATIONS OF THE GLOBAL DATABASE

The global database represents the problem states, and in the town example this database consists of four tables.

1. JUNCTION [J#: Density-of-Houses]

Each junction is individually identified by a number (J#), with a property (Density-of-Houses).

2. JUNCTION-EXIT [J#, E#: R#, Direction]

At each junction (J#) there is an exit (E#) which sets off in a particular direction along a specific road (R#).

3. ROAD [R#: J#1, E#1, J#2, E#2, Length, Type]

All the junctions are linked together by roads (R#) that have properties (Length and Type) which may be pertinent to the task domain.

4. ROUTE [R#1, R#2: Sequence#:]

Finally, there is the route through the town expressed as a sequence of roads. There are parts of the route that do not constitute elements of the solution because they were 'mistakes'. However, there may be several routes, each being a different solution to a problem. These different routes and 'mistakes' are not kept as a permanent record. Sequence# is an alternative key under these conditions, since a road would not be retraced except in the process of finding the solution. Normally the route would consist of a simple list showing the sequence of roads to be taken, but in the initial stages of analysis this list is reduced to its components. The list may be re-formed by a process of self-join.

The constraints of the database are given in the form of a semantic functional dependency (Sfd) graph in Figure 7.1.

These constraints are maintained through the application of production rules, and the database is updated as more is found out about the town. Initially, all that is known is the nature of the first junction. As the strategy of getting through the town is employed, a 'map' of the town slowly unfolds.

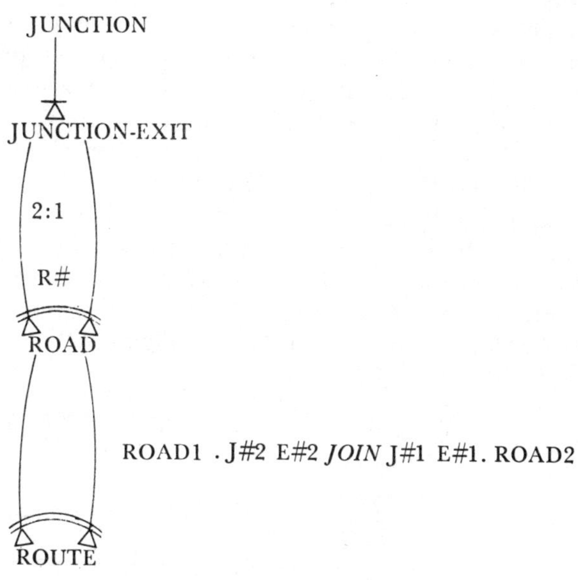

Figure 7.1 *The Sfd graph to contain knowledge of a town*

Every junction must have at least one exit (an exit is also an entrance, and a junction with a single exit represents a cul-de-sac). Each road will be defined by a pair of junctions. In order to ensure that :

1. The roads are linked by a single pair of J#E#, then for each R# there are two and only two J#E#. This is indicated by the enclosed 2:1.
2. The roads are linked coherently then the ROAD is formed only through the join of JUNCTION-EXIT with itself over R#. It is in this way that the map unfolds by a form of inference.

The sequence of roads that make up the route must also be coherent so that no 'jumps' are possible. This is achieved through the self-join of ROAD such that the second reference to J# (J#2) in the first reference to ROAD (ROAD1) is joined over the first reference to J# (J#1) in the second reference to ROAD (ROAD2). Similarly, and at the same time, the self-join is done with E# (the exits belonging to the junctions). This operation is indicated by the modified join symbol where the first element in the prefix dot list joins with the first element in the postfix dot list and so on.

To represent the global database in the form of a set of tables (as suggested by the analysis) might not be the best solution (Addis, 1982c). Representations in main memory can be kept efficiently in different structures. Arrays, lists and networks contain their own natural constraints that may match the problem under consideration. For example, the 'tiles problem', in its simplest variation, is a 3 X 3 square grid where eight of the nine positions each contains a numbered tile. The tiles cannot be removed from the grid, but they can slide within the grid frame along a row or up or down a column provided there is a vacant space. The task is to find the least number of tile moves which will change the existing configuration of tiles into a diffeı ent configuration. There are 362,880 possible different states. Figure 7.2 illustrates the problem.

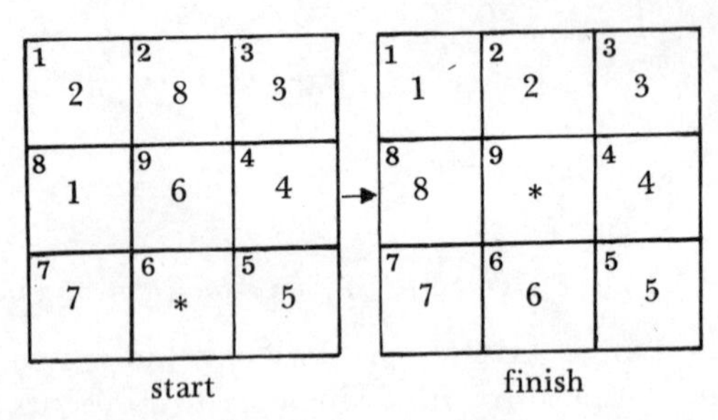

* = space

Figure 7.2 *The eight-tile puzzle (adapted from Nilsson, 1980)*

An analysis of this problem recognizes seven tables:

1. GRID [Position:]

   gives all the grid position numbers.

2. GRIDTILE [Position:TileNumber]

   shows how the tiles (TileNumber) are distributed over the grid (Position).

3. TILEGRID [TileNumber:Position]

   is a complement to GRIDTILE, and indicates where the tiles are positioned. The information is identical to that of GRIDTILE, and it is only the key that has changed. This table is needed to ensure that a tile does not appear in two positions at once (as it could do if it remains just as an own attribute). It is an alternative key attribute, and as such can be represented as a complementary entity set.

4. TILE [TileNumber:]

   is the set of tiles which may have other properties associated with it (none given but it is possible to indicate the markings on individual tiles as an own attribute, for example, and it may be the configuration of these markings that is important in specifying a goal).

5. SPACES [Position:]

   normally contains only one element but in principle there could be several missing tiles and consequently a greater choice of moves.

6. NEXT-TO [Position, Neighbour: Direction]

   gives the structure of the grid. This representation allows many interesting structures to be specified which are well beyond the simple square. Cylinders, spheres, planes with bridges and tunnels or multidimensional systems may be expressed. The attribute 'Direction' currently has values such as 'left', 'right', 'up' and 'down'. This attribute can be used by the production rules to find a solution.

7. *MOVES [Neighbour, Position: Direction, TileNumber]

   is the list of next possible moves where Direction has been inherited from NEXT-TO. MOVES is not set in Boyce/Codd Normal Form (BCNF) (hence the *) since the own domains are inherited.

Figure 7.3 gives the Sfd graph for the system where # is used to indicate the fixed nature of the tables. Table 7.1 shows some of the tables in the starting state.

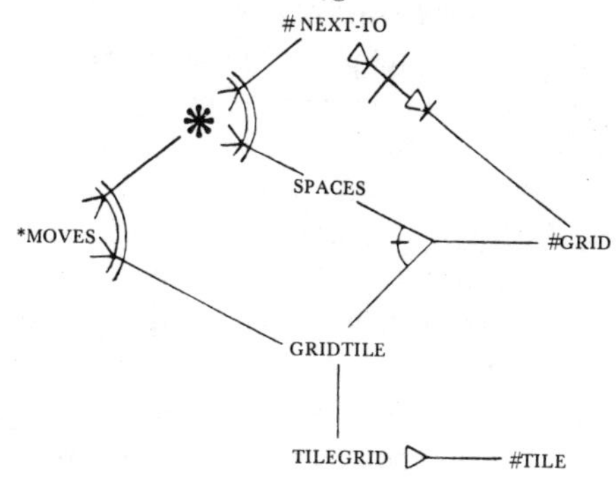

Figure 7.3 *The Sfd graph for the tile puzzle (cf Figure 7.2)*

GRIDTILE
[Position: Tile Number]

Position	Tile Number
1	2
2	8
3	3
4	4
5	5
7	7
8	1
9	6

SPACES [Position:]

Position
6

TILEGRID
[TileNumber: Position]

TileNumber	Position
1	8
2	1
3	3
4	4
5	5
6	9
7	7
8	2

NEXT-TO
[Position Neighbour: Direction]

Position	Neighbour	Direction
1	2	right
1	8	down
2	1	left
2	3	right
2	9	down
3	4	down
3	2	left
4	3	up
4	9	left
4	5	down
5	4	up
5	6	left
6	5	right
6	9	up
6	7	left
7	6	right
7	8	up
8	1	up
8	9	right
8	7	down
9	8	left
9	2	up
9	4	right
9	6	down

MOVES
[Neighbour, Position: Direction, TileNumber]

Neighbour	Position	Direction	TileNumber
5	6	right	5
7	6	left	7
9	6	up	6

Table 7.1 *The starting state for the tile puzzle*

It is clear from this example that the simple grid representation is compact, clear and contains many of the constraints 'built in' naturally. What the grid does not present is a clear illustration of all the potential variations. The art is to analyse the problem into the most freely expressible form, and then to choose a representation scheme that fits the properties of both the problem and the implementation. *Ad hoc* schemes very rarely have either a fixed set of well-understood operations or display all the degrees of freedom explicitly. The use, size and data distribution would govern the final implementation details (see Chapter 4), but for the purposes of explanation a simple table model will be assumed.

PRODUCTION RULES

Production rules are the set of transformations that may be applied to a representation scheme. They essentially have two parts: a contextual part that expresses when a rule can be used, and a transformational part that shows what changes take place when the rule is applied. Some rules (sometimes called demons) are obligatory and must be applied when the contextual part matches the situation. Other rules (sometimes called servants) are optional and are open to selection by the control strategy. In general, the rules concerned with maintaining constraints are obligatory, and those rules involved with the problem solution are optional.

The tiles problem may be defined by expressing the constraints as a series of rules which are interpreted as *if* situation *then* apply a transformation. The obligatory rules are derived from the Sfd graph (see Figure 7.3). Examples of their form (where words set in italics are reserved words) can be presented using relational operations.

The symbol ':=' is used to mean 'make equal to', whereas the symbol '=' is the Boolean operation 'equals' and results in 'True' or 'False'. Note that the projection symbols [ ] are used to alter the order of the column headings (and hence the columns are re-ordered as well) so that two tables may be compared. The column names refer to the names given in the original definition of the tables (see p. 158). The rules are mandatory or optional.

Mandatory rules

MR1 *IF NOT* TILEGRID [TileNumber] = TILE
THEN PRINT ('Error') *FI*

MR2 *IF NOT* TILEGRID = GRIDTILE [TileNumber, Position]
THEN TILEGRID:=GRIDTILE [TileNumber, Position] *FI*

MR3 *IF NOT* GRID *MINUS* GRIDTILE [Position] = SPACES
THEN SPACES:= GRID *MINUS* GRIDTILE [Position];
MOVES:= (NEXT-TO *JOIN* SPACES)
.Neighbour *JOIN* Position.GRIDTILE *FI*

Optional rules

```
OR4    IF NOT (GRIDTILE=GOAL) AND PRESENT (MOVES WHER
       Direction = 'left')
       THEN GRIDTILE:= GRIDTILE MINUS
                       (GRIDTILE JOIN ((MOVES WHERE
                       Direction = 'left') [TileNumber]))
                       UNION
                       (GRIDTILE JOIN
                       ((MOVES WHERE Direction = 'left')
                            [Position]));
                       PRINT ('MOVE' (MOVES WHERE
                       Direction = 'left') [TileNumber]
                       'left') FI
```

OR5, OR6 and OR7 are identical to OR4 except that for 'left' read 'up', 'right' and 'down', respectively. GOAL is a relation giving the desired pattern of tiles to be achieved. These optional rules can be expressed as a single process with a parameter (direction), and the choice of parameters is made at each stage from MOVES. A further improvement can be made by selectively merging with this parameterized rule those mandatory rules that are known to be activated by the consequence of applying optional rules. However, this will reduce the generality of the system to cope with the addition of new constraints or an extension to other problems.

The rules, in this case, all represent 'negative feedback' in that they attempt to maintain the database by testing for deviations from a given norm. The rules react to the deviations by transformations that reduce the 'displacement'. MR1 has no means of correction other than an appeal to the programmer (error message). MR2 checks the bijective link between TILEGRID and GRIDTILE (see Figure 7.3). The rules MR1 and MR2 are used only on the initial setting up of the starting situation, and they will not normally be triggered during the search for a selection. The response is chosen by assuming GRIDTILE as the desired pattern, since it is through OR4, OR5, OR6 and OR7 that GRIDTILE will be changed. MR3 ensures that the SPACES in the GRID (where there are no tiles) are recorded correctly, and it extracts all the latest available moves if different from those recorded. All mandatory rules must be applied (if applicable), since they govern the constraints of the system. The optional rules can be selected (if applicable) since they describe the choice of transformations towards a given solution. These rules are also 'negative feedback' rules in that they detect a difference between the desired state (goal) and the current state, then react in a way to reduce the difference. How this choice is made and the manner in which the difference is measured depend upon the control strategy and the heuristics.

CONTROL STRATEGIES

The control strategy is concerned primarily with optional rules. The system, as described, would enter into a stable loop unless some direction was given. One method is to assign a measure of 'closeness' to the goal that can be calculated from the present state. Such a measure is called either the 'state-goal function' or the 'evaluation function'. The state-goal function has the property of being at its maximum value when the goal is achieved, whereas the evaluation function is an assessment of the *cost* of achieving a goal.

Irrevocable Strategies

Figure 7.4 shows the classes of control strategies that can be used for rule selection. Irrevocable control strategies simply choose a rule and apply it, and once the transformation has been done it is never explicitly reversed. Such a strategy can only be used provided that there is either infallible local knowledge or the order of applying the rules is irrelevant. Infallible local knowledge refers to the state-goal function or an equivalent method of choosing the correct rule at each stage in determining the solution. A common approach is to use a 'hill climbing process' that selects a rule that will either increase the state-goal function or at least maintain its current value. (This can be visualized as being equivalent to climbing a hill in a fog hence the term 'hill climbing').

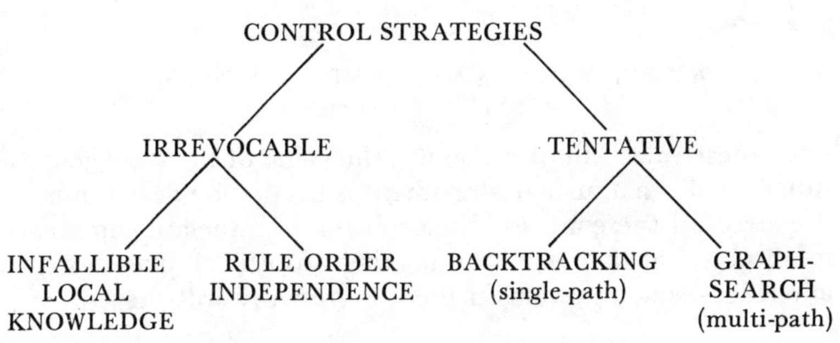

Figure 7.4 *Classes of control strategies*

The state-goal function is a heuristic device in that it is derived through an insight of the problem domain. In the tiles problem a suitable function might simply be 'minus the number of tiles out of place'. Thus, if the initial pattern is:

2	8	3
1	6	4
7	*	5

and the goal pattern is:

1	2	3
8	*	4
7	6	5

then the state-goal function value will be -4 for the initial pattern and zero for the goal pattern. The rules can then be chosen according to the best 'increase' in this value, thus generating a sequence (state-goal values are set in italics):

– 4		*– 3*		*– 3*		*– 2*		*– 1*		*0*
2 8 3		2 8 3		2 * 3		* 2 3		1 2 3		1 2 3
1 6 4	up →	1 * 4	up →	1 8 4	left →	1 8 4	down →	* 8 4	right →	8 * 4
7 * 5		7 6 5		7 6 5		7 6 8		7 6 8		7 6 5

As can be seen from Table 7.1 a pre-emptive heuristic might be:

TRY RULES WHERE NOT MOVES [Tile Position]
= MOVES [TileNumber]

since these rules will often reduce the value of the state-goal function. Then a further assessment is needed by determining the effect of these moves. The problem with the hill climbing method is that there are local maxima (foothills), plateaux and ridges which prevent further progress towards the solution. For example, if the initial state is:

	−2	
1	2	5
*	7	4
8	6	3

then any move will decrease the state-goal function value.

Tentative Strategies

Tentative strategies provide the opportunity to explore the problem space and to determine a solution from the experience. The 'town problem' (see Figure 7.1) illustrates the need to backtrack from cul-de-sacs or routes that are definitely leading away from the desired objective. As more is discovered about the town, the possibility of finding a path through the town is improved. Consequently, it is important to record the information gained during these manoeuvres.

Backtracking, in its basic form, does not use a state-goal function, since the principle depends upon determining local knowledge empirically. Control rules that govern the manner in which the problem space is explored are required. These rules determine:

1. The selection of a move (transformation or optional rule);
2. The conditions under which exploration ceases and backtracking takes place.

In the example of the tiles problem, the rules can be scanned in the order: left, up, right, down (say). The backtrack conditions may be:

1. The same state of the global database has appeared in this solution path as has appeared earlier. This, of course, means that the path has doubled back on itself, and therefore another choice is needed.
2. The path travelled has gone beyond a certain number of moves (depth bound). This embodies some knowledge of the expected 'size' of the solution beyond which it is not reasonable to go (say six for the tiles problem).
3. There are no more optional rules to apply, and therefore it is necessary to backtrack to an earlier state where there was a choice.

The example of the tiles problem (from Nilsson, 1980) using these rules is shown in Figure 7.5. Only one path is kept, and all decisions are made from this path. The other information retained is the set of rules still untried at each stage in the path.

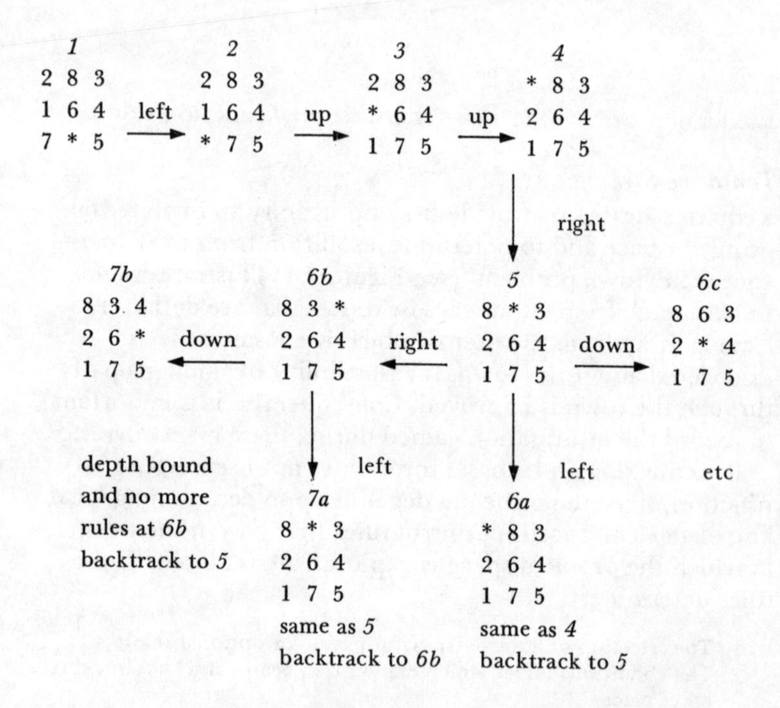

Figure 7.5. *The backtracking strategy example of the tiles problem (adapted from Nilsson, 1980)*

The problems with backtracking are that it may generate non-terminal databases forever, that it may cycle and that it may fail due to a move made very early in the exploration. Cycles may be avoided by a depth bound search; and a better failure analysis would alleviate the difficulties caused by an early false move. Graph search, on the other hand, explores many paths simultaneously, and is of particular interest because it encompasses many of the elements of other strategies.

Graph search

A problem space can be represented as a directed graph where each node is a particular state of the global database. Each directed arc that connects the nodes is a transformation applied to one state and converts it into another. In the case of the town: the problem space is topologically equivalent to the road map of the town, the junctions are the states and the connecting roads are the transformations. The arcs and nodes of the problem space are then set within the three-dimensional reference space of the compass points, the distance and the house density. The tiles problem space consists of: nodes showing the distribution of tiles within the grid, and arcs indicating the moves that take one pattern of tiles to the next. Figure 7.6 shows a small segment of the tiles problem space, and an examination of symmetries of such problem spaces can produce elegant control strategies.

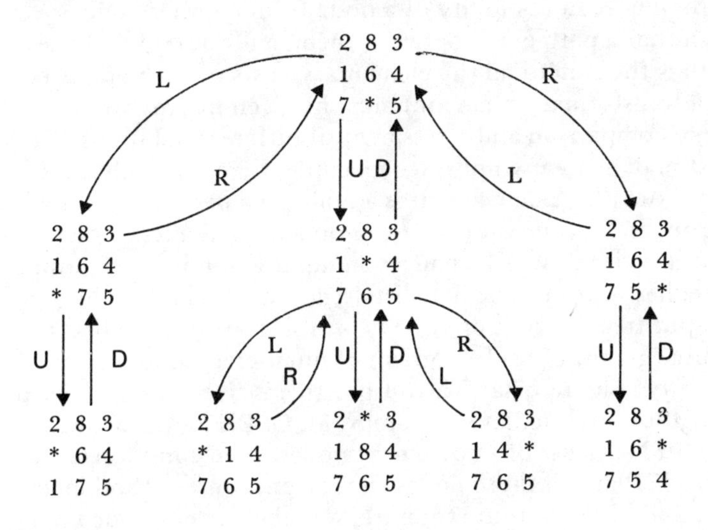

Figure 7.6 *A segment of the tiles problem space*

In practice, most problem spaces are too large to store within a computer, and it is not necessary to have all the problem space since the solution is only concerned with a small part. Itineraries given for holiday travel only plot 'the solution' ignoring most of the surrounding terrain, and it is desirable, in the 'town problem', to find the shortest route with minimum effort. What would be better is some surety that the route *is* the shortest. One result from graph search theory is that such an assurance is given provided certain conditions are met.

A GENERAL GRAPH SEARCHING ALGORITHM 'A'

The general graph searching algorithm explores a directed graph, and traces out a tree-structured path. Such a path has certain properties that are exploited by the control strategy. The properties of depth, ancestor, successor, accessibility and path length can be clearly defined for a tree.

A tree is a graph where each node has at most one parent. The root node has no parents, and is at depth zero. A tip (or leaf) node has no successors. The depth of a node is defined recursively as depth of parent node plus 1, and a path length is the depth of the end minus the depth of the start of the path plus 1. Accessibility of a node from another node is such that a path exists between them, and the 'cost' of the path is the sum of all the elementary costs of each arc. A cost of a transformation (an arc) must be given in order to make some comparison and assessment of different solutions. The cost may be in any units, for example, time, difficulty, distance or risk. A descendant is a node on a path at a greater depth, and an ancestor is a node on a path at a lesser depth. The algorithm, which changes an input graph into an output tree, has four intermediate lists. The final 'shape' of the output tree depends upon the control strategy as well as the source graph. Consider a graph of numbered nodes as illustrated in Figure 7.7(a). The output tree is given in Figure 7.7(b). The four intermediate lists are N, M, OPEN and CLOSED.

OPEN is a list of known next nodes to be considered, CLOSED are all those nodes already encountered and incorporated in the output structure, N is the current node under consideration and M is the 'expansion' of the nodes in N less those in OPEN or CLOSED. An expansion of a node is the set of neighbouring nodes produced when all the applicable transformations have been employed. The elimination of those nodes on OPEN and CLOSED from M ensures that a node is never revisited.

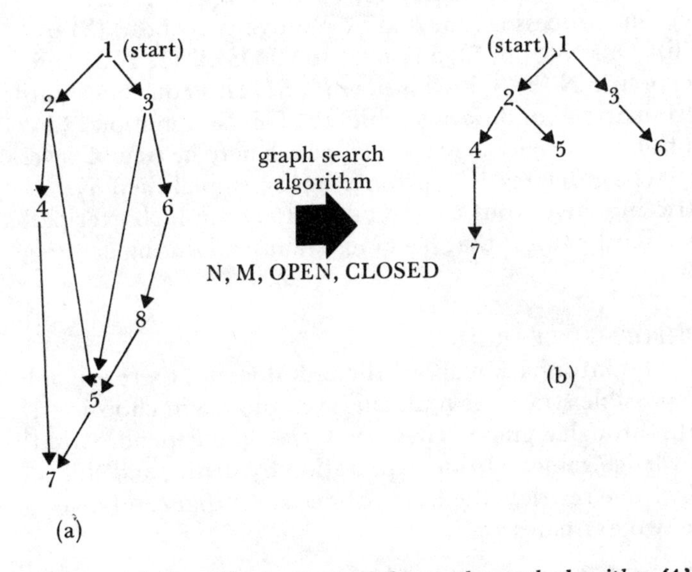

Figure 7.7 *The input and output of the graph search algorithm 'A' (adapted from Nilsson, 1980)*

The process has the first node in OPEN (1) (the contents of each list are given in parentheses). It is then removed from OPEN, and placed in both CLOSED and N. The node is then expanded (the generation of possible next steps) into M (2, 3), and since in this initial stage there are no other nodes visited this expansion is placed directly into OPEN (2, 3). The root of the tree is set up in the output (1) with the two branches (2, 3).

Next, an important activity takes place. The nodes (or at least their representatives) in OPEN are sorted according to some principle (the tests). This sorting is determined by the control strategy, and governs which node is to be explored next. Details of this part will be discussed more fully later. For the purposes of this description the new nodes will always be put on the tail of the list.

OPEN (2, 3) is reduced by removing the head (2), the head is added to CLOSED (1, 2) and is also placed in N (2). This node (2) is then expanded (4, 5), and any nodes in OPEN (3) and CLOSED (1, 2) are removed before placing them in M. M (4, 5) is added to the tail of OPEN (3, 4, 5), and the new branches (4, 5) of the growing tree are spliced into the node (2). The sequence in OPEN is sorted (in this case no sort). The process is repeated by removing the head (3) of the list OPEN (3, 4, 5), adding it to CLOSED (1, 2, 3) and placing it in N (3). OPEN is now (4, 5). This continues until a termination condition is achieved. This termination condition is usually a goal node, which may be one of several possible candidates. The process can be embellished by restricting the output tree to take into account shorter paths (lower cost routes) that become established during exploration.

UNINFORMED SEARCH

Given no information about the task domain, there are only two possible strategies that can be employed in choosing a route through a graph. These two strategies depend upon the knowledge gained during exploration by depth, and this knowledge restricts the kinds of decisions that can be made. The two extremes are:

1. *Depth-first* takes a path down a tree such that the route chosen is always deeper or until a tip node is reached. If this is not a goal node, then it will backtrack one node and try a different branch. Depth-first also applies to any directed graph using algorithm 'A'. When using algorithm 'A', the list in OPEN is sorted according to the property of depth descending (greatest depth first). The sorting process can be avoided by simply appending each new expansion in M on the *front* of the list in OPEN.
2. *Breadth-first* explores every node at one level before descending to the next level. In algorithm 'A' the list in OPEN is sorted according to depth increasing (smallest depth first), and this is achieved by appending each new expansion in M to the *end* of the list in OPEN. The example describing 'A' (see p. 168) was breadth-first. Breadth-first has the important characteristic that it will find the shortest path to a goal if it exists. This is because it makes sure, at each level, that no goal node exists before descending.

Depth-first may well find a termination condition for less cost than breadth-first, but there is no guarantee that this termination condition is the least expensive. The cost of finding a good solution has to be balanced against the cost of the found solution. There is also no guarantee that depth-first will ever find a solution or do so at less cost than breadth-first. However, with many of the simple problems with finite problem spaces, depth-first is usually the cheapest strategy.

What is needed is a method which gives the best solution at minimum cost. This requires some knowledge of the task domain.

TASK DOMAIN KNOWLEDGE FOR SEARCH
Knowledge for guiding the search algorithm 'A' is usually embodied in the heuristic 'evaluation function'. This function is an estimate of the total 'cost' of a solution that may be found through a given node. There are other kinds of heuristic knowledge that may come in the form of instructions but only the evaluation function will be considered.

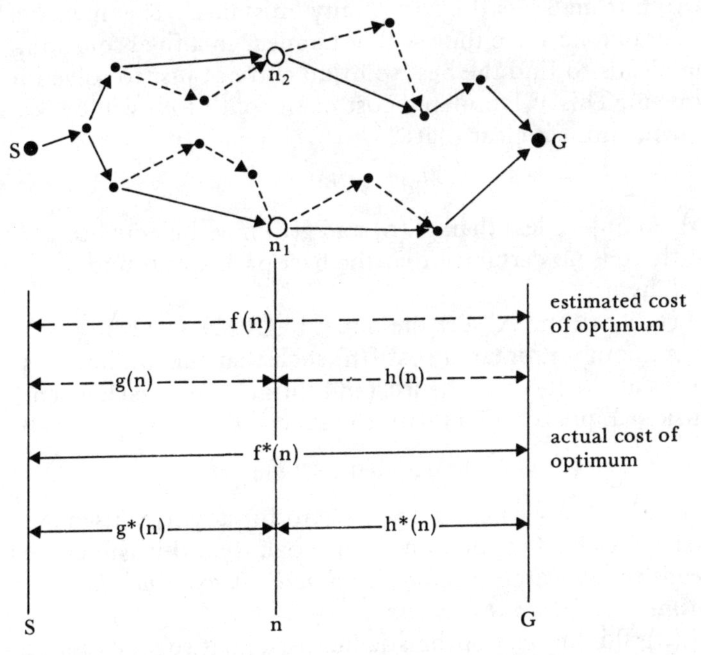

Figure 7.8 *The evaluation function f(n) of optimum path through n*

Figure 7.8 shows a possible 'optimum' solution path passing through node n1 and another 'optimum' solution path passing through n2. Each of these paths is optimum in the sense that no shorter (lower cost) path exists between S and G if the path is forced to pass through the given node n. There may be longer (more expensive) paths that contain each of these nodes. The optimum solution path through a node n will be expressed as f*(n). This path can be divided into g*(n) which is the optimum path from S to n and h*(n) which is the optimum path from n to G. The problem is to determine a method of estimation that will ensure that the optimum path is always traversed by algorithm 'A' with minimum backtracking.

The estimation naturally falls into the two parts g(n) which is an estimate of the cost of getting to n ignoring all retracing of steps and h(n) which is an estimate of the cost of reaching G from n. Thus:

$$f(n)=g(n)+h(n) \tag{7.1}$$

Now g(n) is simply the sum of the cost of each arc traced to get to n from S less the cost of any 'mistakes'. The 'mistakes' are eliminated since they will not appear in a final solution. The idea is to find the best solution and not just to solve the problem. This accumulated cost of the path to n will be known, and it is clear that:

$$g(n) \geq g^*(n) \tag{7.2}$$

g(n) cannot be less than g*(n) and g(n) may be equal to g*(n) but there is no certainty that the best path was found in reaching n.

Let algorithm 'A' sort the nodes in OPEN according to the minimum estimated cost f(n) such that the minimum is always at the head of the list (and thus chosen first). Then consider Equation (7.1) with the special case where h(n)=0:

$$f(n)=g(n) \text{ where } h(n)=0 \text{ at all times}$$

Here if h(n)=0 [ie h(n) is always zero for any n] is used as the basis of cost estimation to the goal, then this will cause a breadth-first search. *Such a search will always find the optimum solution if it exists.*

h(n)=0 is an over-optimistic heuristic that suggests that it will cost nothing to achieve the goal from n. The inverse of this is to make g(n)=0, set h(n) to some large number and reduce h(n) by 1 for each level explored. This will result in a depth-first search with no certainty that an optimum path has been found.

A balance between these two extremes is required. An alternative reference is the actual cost of h(n), namely h*(n). It can be shown that provided h(n) is always an optimistic estimate such that:

$$h(n) \leq h^*(n) \tag{7.3}$$

then *an optimum path will always be found*. The algorithm 'A' that uses such a heuristic for control is called 'A*'. A* is called ADMISSIBLE because it finds the shortest path. The reason for this can be illustrated by considering initially the extreme case of h(n)=0 at all times. Under these conditions, g(n)=g*(n) because a breadth-first search will always find the optimum path to any n. If h(n) is always an underestimate of the cost (optimistic) to a goal from n and we assume that the cost to reach n is a minimum, then:

$$f(n)=g^*(n)+h(n) \tag{7.4}$$

but $h(n) \leq h^*(n)$. Therefore:

$$f(n) \leq g^*(n)+h^*(n) \tag{7.5}$$

but:

$$f^*(n)=g^*(n)+h^*(n) \tag{7.6}$$

and is the minimum cost. Therefore, from Equations (7.5) and (7.6):

$$f(n)=f^*(n) \tag{7.7}$$

The assumption that the cost to reach n is minimum can be dispensed with since this cost is known to be zero at (the start) S. Since this argument applies at every step away from S to G, then the optimum path is known to be found up to any node (n) because of Equation (7.7). Therefore, the assumption that g(n)=g*(n) is indeed correct. This argument and the admissibility of A* do not depend upon the 'monotone restriction'. The monotone restriction is simply that the estimate h(n) will always reduce for each new estimate of a node that is closer to the goal.

Under these stronger conditions, it is known that A* will always select an n such that:

$$g(n)=g^*(n)$$

and thus if the restriction holds, A* will find the optimum path with much less backtracking than if it does not hold. Note that g*(n) is the optimum path *to n* and is not necessarily *on the optimum path to G* (unless G=n).

The nearer h(n) is to h*(n), the closer the total path traced (including errors) is to optimum. This result is important because heuristics that are devised to give h(n) can be judged in many cases to be underestimates or overestimates relatively easily. If many of the constraints of the problem are ignored, then for certain problems this will tend to produce an optimistic estimate because the excess moves to 'get round' these constraints are ignored. For example, in the tiles problem the estimate ignores the effect of moving a tile or even how far a tile is away from its required position. An improved heuristic would be:

h(n)=sum of tile displacements

In a complete design, the argument for the estimation power of h(n) must be given since this will provide important information as to why it was chosen, suggest the potential performance of the completed system and give a source for improvement.

Decomposable production systems

There are some tasks that can be decomposed into sub-tasks of a similar nature to the main task. A solution is the concatenation of all the sub-solutions. For example, consider the problem of a boy who collects stamps (M). He has for the purposes of exchange a winning conker (C), a bat (B) and a small toy animal (A). In his class there are friends who are also keen collectors of different items and will make the following exchanges:

1. 1 winning conker (C) for a comic (D) and a bag of sweets (S);
2. 1 winning conker (C) for a bat (B) and a stamp (M);
3. 1 bat (B) for two stamps (M, M);
4. 1 small toy animal (A) for two bats (B, B) and a stamp (M).

The problem is how should he carry out the exchanges so that all his exchangeable goods are converted into stamps (M). This task can be expressed more briefly as a production system:

(a) Initial State of Database = (C, B, A)
(b) Transformation rules:

1. *IF* C *THEN* (D, S)
2. *IF* C *THEN* (B, M)
3. *IF* B *THEN* (M, M)
4 *IF* A *THEN* (B, B, M)

(c) Goal: to be left with only stamps (M, - - - -, M)

The production system described on p. 174 would do a lot of extra work by redoing many of the transformations. In this example, we could create the breadth-first graph shown in Figure 7.9. As will be seen, much of the effect in expansion is repeated down each limb of the graph.

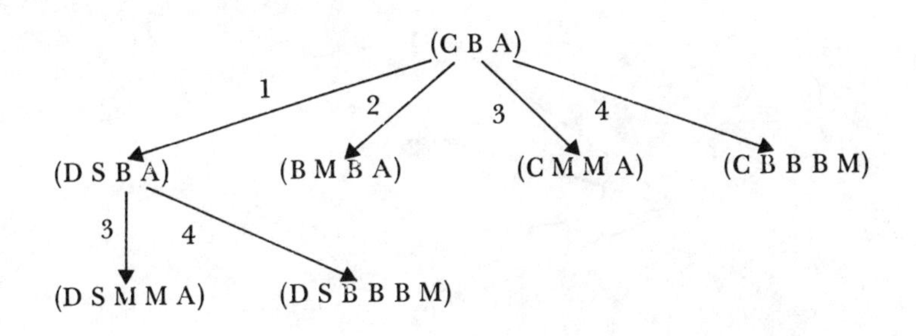

Figure 7.9 *Part of the expansion of the exchange problem (adapted from Nilsson, 1980)*

This repetition can be avoided by decomposing the problem into sub-problems. Thus, the initial problem converts into the sub-problems of how can the boy change his conker into stamps, his bat into stamps and his small toy animal into stamps. This principle can then be reapplied at each stage of the exchanges, and it can be represented by an AND/OR graph.

AND/OR GRAPHS

The AND/OR graph is a specialization of a hypergraph. A hypergraph generalizes the arc from a simple one-to-one directed connector to an m-to-n directed connector, so that m nodes are simultaneously linked to n nodes through one hyperarc. The specialization of the AND/OR graph limits the hyperarcs to k-connectors. A k-connector is a one-to-k link between a single node (n0) and k other nodes (n1 ----- nk). An interpretation of the k-connector is an AND of all the nodes n1 to nk which constitute the node n0. The k-connector is represented as a fan of arrows with a single tie as in Figure 7.10. The single one-to-one arc would be called an 1-connector.

The cost of an arc has been assumed to be some arbitrary unit. It is normally assumed that the cost of a k-connector will be k units. The k-connector is such that it is always thought of as a single entity, and a path involving k-connectors must have every l to k nodes on the path (by definition). Thus, in the exchange example the k-connector can be used for every decomposed part of the problem as shown in Figure 7.11. The different possibilities are indicated by the alternative l-connectors.

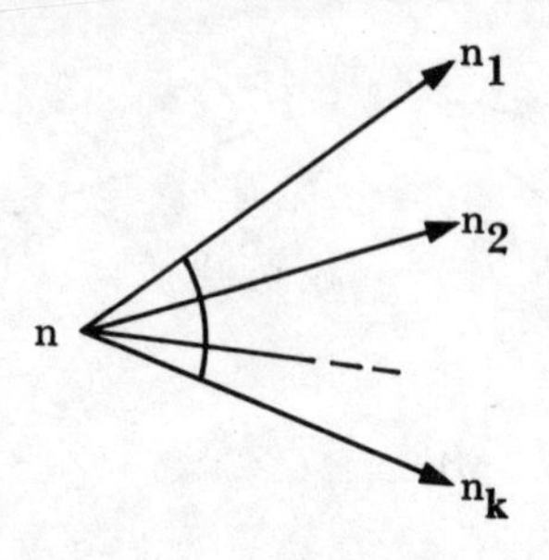

Figure 7.10 *A k-connector*

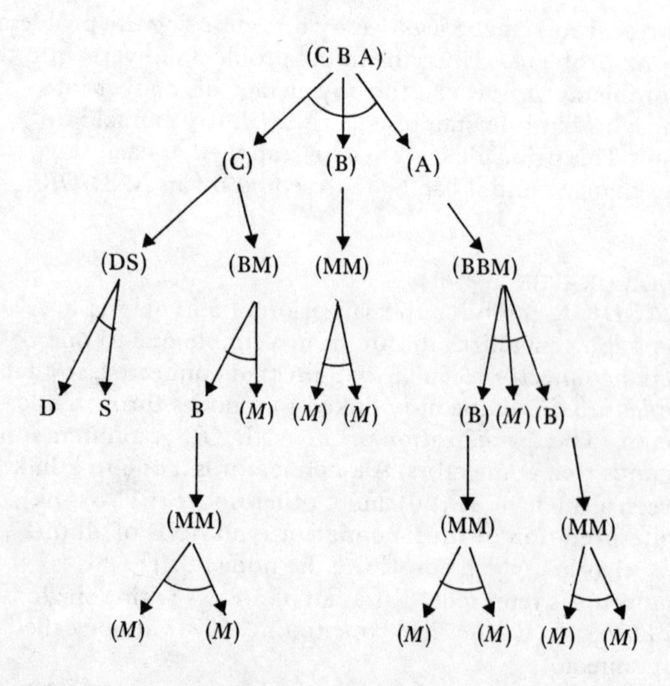

Figure 7.11 *The exchange problem as an AND/OR graph (adapted from Nilsson, 1980)*

The solution for the exchange problem will be:

> **'Swap conker for bat and a stamp, then exchange this bat for two stamps. Swap his own bat for two more stamps, and finally swap the small toy animal for two bats and a stamp .The two bats can then each be exchanged for two stamps'. (The 'cost' would be 22 units.)**

AND/OR HEURISTIC SEARCH

There is a general AND/OR graph searching algorithm AO* that assumes both that the acyclicity (no cycles or loops) and the monotonic restrictions hold. The objective of AO* is to determine an optimum path involving k-connectors from a single start node (n0) to a set of goal nodes where the conjunction of the elements of this set constitutes a single solution. The example shown in Figure 7.12 (from Nilsson, 1980) has a collection of nodes limited by k-connectors. The nodes have pre-assigned h-values that give an estimate of the cost of achieving a solution from that node. The cost of traversing a k-connector is k, and the assessment of the cost is adjusted as more information is gathered during the search. The final solution is the path that accounts for all the nodes of every k-connector and involves the lowest cost.

Figure 7.12 (1 to 5) illustrates the stages of tracing a path from node (a) to solution nodes (i j). Initially the heuristic's estimate of the cost to solution at (a) is over-optimistic with h(a)=0. Both possibilities are explored (b and e d). The estimate of h(b)=2 plus the cost of a single arc (+1) is the estimated cost to solution down arc (a b), whereas the estimate h(e)=1 and h(d)=1 plus a 2-connector gives a total estimated cost of 4 down arc (a(e d)). Thus the preferred route is from (a) to (b), with the new estimate of the cost at (a) now being equal to 3 [ie q/(a)=3 where q is the update of h at each node].

Expanding (b) to (f) or (c) increases the cost at (b) to 5 [ie q/(b)=5], therefore, the cheapest estimate is now from (a) to (e d) where q/(a) becomes 4.

The node (e) is expanded to either (g) or (i j). The latter is the desired goal, therefore node (e) can be marked as a sub-goal and updated with the known cost. Node (d) is now expanded to (e) or (i). The total cost of the solution:

(a) to [(e) to (j i), (d) to (i)]

is 5 units.

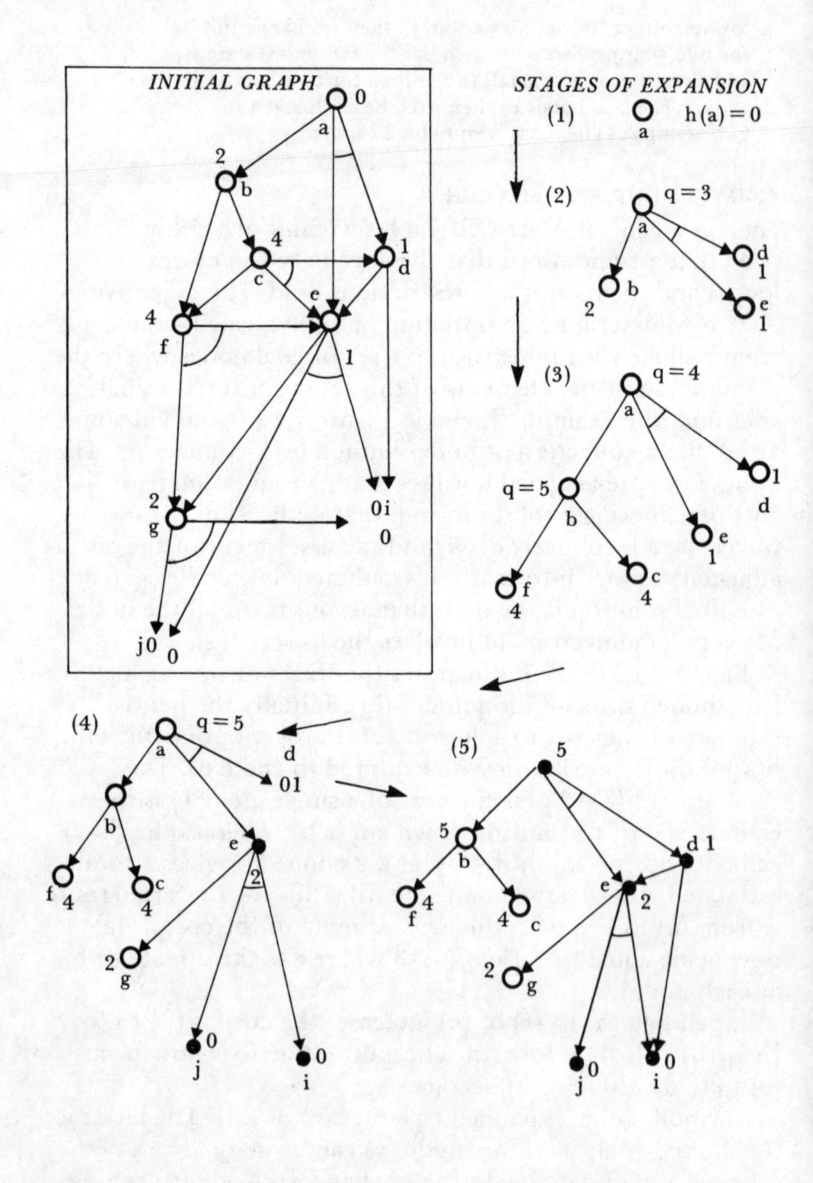

Figure 7.12 *An example of AO* heuristic search (from Nilsson, 1980)*

The AND/OR graph is used extensively in rule-based inference. More examples will be given in Chapter 8 where mechanical reasoning will be examined.

CHAPTER 8

Mechanical Reasoning

> The rabbit brain has evolved to suit its way of life, beset with many hazards, a primitive brain which translates the messages from its well-developed sense organs into what seems to us at times action of a reasoning kind, though at other times the rabbit may appear to behave in a stupid fashion.
>
> *R.M. Lockley (1964)*
> *The Private Life of the Rabbit*
> *with kind permission of Andre Deutsch*

Introduction

The claim that the essence of intellectual ability can be captured within a computer program requires that this ability can be described, at least in principle, as a formally expressed rubric. However, a distinction should be made between naturally occurring performance and its representation within a formal system. For example, the moon's orbit round the earth can be described within a formal system of mathematics and simulated within the computer, but there is in this simulation neither the suggestion that the moon 'calculates' its position with respect to time nor that there are necessarily any physical representations for such calculations (Dreyfus, 1972). A similar distinction should be made between human behaviour and artificial intelligence programs. Certain kinds of restricted language activity (such as the protocol analysis of Newell and Simon) can be monitored and described, but there is rarely any conclusive evidence that there are physiological or psychological correlates to these descriptions (Gough, 1966).

The outward appearance of reason is 'the argument' expressed as a set of connected sentences. Each step in the argument can be accepted as 'correct' according to agreed principles. The normal debate between two rational people involves sentences too complex and too subtle to engender in a computer program (at least, some believe, just for the moment). Yet artificial intelligence, if it is to be recognized as a process of reasoning, can hardly be said to take place unless there is an explicit manifestation of argumentation.

The production rule system described in Chapter 7 fails in this respect in that no reference is made to 'arguing the case'. There are, however, formal systems that manipulate *standardized* sentences according to well-specified rules and allow certain kinds of inferences to be made. Such a system is the combination of Propositional and first-order Predicate Calculus. A detailed summary of this calculus with an emphasis on its mechanistic character follows. There is also a clarification of the boundary between the calculus and its interpretation.

A calculus of reason

PROPOSITIONAL CALCULUS

Propositional calculus is concerned with complete statements of facts (propositions) such as:

'John loves Mary'

or:

'Birds fly south in Autumn'

The only interpretation required of these statements is that they are either True or False (but not both). A proposition P is a declarative sentence which can be used to produce a grammatical question from the transformation:

'Is it True that P?'

Thus, questions, commands, invitations and ungrammatical sentences are not propositions. How questions based upon the above may be answered is not defined since the assignment of the property True or False to the initial atomic 'fact' is not part of the calculus. The signs True or False can be different (eg 0 or 1), and they do not even need to represent the meaning of the symbol 'true' or 'false' where the normal interpretation may be assumed. (NB A 'sign' is the outward manifestation of a 'symbol'. The same 'sign' can be common to different 'symbols'.)

Propositions may be linked together to form 'expressions' (or compound propositions) using logical connectives. The logical connectives are shown in Table 8.1.

PRECEDENCES	SIGN	MEANING
1	$\sim$	*Not* or Negation.
2	$\wedge$	*And* or Conjunction.
3	$\vee$	*Or* or Disjunction.
4	$\rightarrow$	*If . . . Then . . .* or Implication.
5	$\leftrightarrow$	*If and only if . . . Then . . .* (iff) or Double Implication.

Table 8.1 *The logical connectives*

The assignment of True or False to a formula constructed from atomic facts or formulae of known truth values (the assignment of True or False) is achieved by simply 'looking it up' in a truth table. The table is predefined, and it expresses every possible combination of truth value assignment for unary and binary compound propositions. Hence the truth table is a way of expressing the truth functions $\wedge$, $\vee$ etc. Table 8.2 is the truth table for the logical connectives. 'P' and 'Q' represent arbitrary propositions or expressions in this table.

P	Q	$\sim P$	$P \wedge Q$	$P \vee Q$	$P \rightarrow Q$	$P \leftrightarrow Q$
T	T	F	T	T	T	T
T	F	F	F	T	F	F
F	T	T	F	T	T	F
F	F	T	F	F	T	T
T = True				F = False		

Table 8.2 *The truth table for logical connectives*

Expressions are constructed according to a simple grammar, and expressions that conform to this grammar are called 'well-formed formulae' (wffs). Parentheses are used to make clear the order of truth value assignment, if not otherwise clear. A wff is either a proposition or will have one of the constructions shown in Table 8.3.

wff
$(\sim P)$
$(P \land Q)$
$(P \lor Q)$
$(P \rightarrow Q)$
$(P \leftrightarrow Q)$

Table 8.3 *The 'allowed' constructors for well-formed formulae where P and Q are well-formed formulae (wffs)*

The connectives have an order of precedence that allows the elimination of some parentheses (see Table 8.1). The $\sim$ is usually bound (refers to) to the immediately following wff, then the $\land$ links the pair of wffs on either side. After that, the $\lor$ is determined before the implication sign. In illustrating the order of precedence we have the following equivalent expressions:

1. $\sim P \land Q \lor R \rightarrow A$ is equivalent to $((((\sim P) \land Q) \lor R) \rightarrow A)$
2. $A \rightarrow R \lor Q \land \sim P$ is equivalent to $(A \rightarrow (R \lor (Q \land (\sim P))))$
3. $P \land R \lor Q \land A$ is equivalent to $((P \land R) \lor (Q \land A))$
4. $P \lor R \land Q \lor A$ is equivalent to $((P \lor (R \land Q)) \lor A)$ or $(P \lor ((R \land Q) \lor A))$

The connectives tend to have a similar meaning to that of their natural language counterparts, thus if:

'John loves Mary' is True (P)

then:

'John does *not* love Mary' is False ($\sim$ P)

and if:

'Birds fly south in Autumn' is True (Q)

then:

'John loves Mary *and* Birds fly south in Autumn'

($P \land Q$)is a compound proposition that is True (from Table 8.2).

However, care must be taken in the translation of the 'implication' since it may be used to suggest a conditional connection that has an underlying 'cause'. Thus:

'*If* it is raining *then* the roads are wet'

is certainly True if at least the components are True. However it can also be True that:

'*If* 2 + 2 = 5 *then* black is white'

and:

'*If* John loves Mary *then* the roads are wet'

INTERPRETATION
The 'implication' indicates that interpretation needs both to be formally defined and to take into account a changing environment. John may not always love Mary, and the roads are not always wet. An *interpretation* of a wff is an assignment of truth values to the component atomic propositions. Thus in the statement:

'*If* it is raining *then* the roads are wet'

may have the *interpretations* shown in Table 8.4.

interpretation number	it is raining (P)	the roads are wet (Q)	wff $P \rightarrow Q$
1	T	T	T
2	T	F	F
3	F	T	T
4	F	F	T

Table 8.4 *Different interpretations of 'If P then Q'*

From Table 8.4, it can be seen that to know that 'the roads are wet' does not indicate that 'it is raining' (interpretation 1 and 3), but when 'it is raining' 'the roads are wet' (interpretation 1 is the condition where 'cause' is reflected). Thus, the behaviour of the compound statement to different interpretations fits an accepted view of where an underlying connection is assumed between two component propositions.

Of the different interpretations (1 to 4), only interpretations 1, 3 and 4 *satisfy* the wff. An interpretation satisfies a wff only if the interpretation evaluates to True according to the truth tables. Each interpretation that satisfies the wff is called a *model*.

In Propositional Calculus, it is feasible to enumerate all the interpretations of a wff. Thus, if there are n distinct atomic propositions connected together to form a wff, then there will be two possible interpretations. A wff is called *valid* only if it is True under all its possible interpretations (otherwise it is *invalid*), and it is unsatisfiable (or inconsistent) only if it is False under all its possible interpretations (otherwise it is satisfiable or consistent). The compound proposition 1 is neither valid nor unsatisfiable. It is, however, invalid and satisfiable, and, consequently, it defines a set of models that may be judged as representing a perception of the environment. It says something about the world (or at least the set of models does).

TAUTOLOGIES, LOGICAL EQUIVALENCE AND VALID ARGUMENTS

Tautologies are, in general, any valid wff. Two formulae P and Q are logically equivalent, $P \equiv Q$, if and only if $P \leftrightarrow Q$ is valid. $P \equiv Q$ has the property that in any other wff, P can replace Q and Q can replace P without any change in the final truth condition for *all* interpretations. P and Q are normally compound propositions.

A convenient way of viewing logical equivalence of this kind is as a set of production rules (that combine in one rule both forward and backward rules) for solving problems in Propositional Calculus. To prove an expression Q (the goal) from a single given wff (the initial database), a process of selecting a sequence of equivalences ending in Q that will create Q from the original wff is required. The equivalences are steps in a compound *argument*, and each step is valid (ie always True no matter what the interpretation). Table 8.5 gives some equivalences with a reference.

An *argument* is a set of wffs (premises) followed by a single wff called the *conclusion*. Arguments can be concatenated into a single compound argument or proof. A complete argument can be made in one statement by 'if ⟨premises⟩ then ⟨conclusion⟩'. Each step in a compound argument (subargument) must be justified, and such a justification is validity. In general an *argument is valid* only if the conjunction of premises (eg P and R) implies the conclusion (eg $P \wedge R \rightarrow Q$ is valid), and this compound proposition is also valid. The validity of an argument is determined only through inspection of all the interpretations. For complex situations, this can be both tedious and time consuming.

reference	logical equivalence	comment
T1	$\sim(\sim P) \equiv P$	double negative or inverse
T2	$P \vee Q \equiv \sim P \rightarrow Q$	also $P \rightarrow Q \equiv \sim P \vee Q$
T3	$\sim(P \wedge Q) \equiv \sim P \vee \sim Q$	De Morgan's Law 1
T4	$\sim(P \vee Q) \equiv \sim P \wedge \sim Q$	De Morgan's Law 2
T5	$P \wedge (Q \vee R) \equiv (P \wedge Q) \vee (P \wedge R)$	Distributive Law 1
T6	$P \vee (Q \wedge R) \equiv (P \vee Q) \wedge (P \vee R)$	Distributive Law 2
T7	$P \wedge R \equiv R \wedge P$	Commutative Law 1
T8	$P \vee R \equiv R \vee P$	Commutative Law 2
T9	$(P \wedge Q) \wedge R \equiv P \wedge (Q \wedge R)$	Associative Law 1
T10	$(P \vee Q) \vee R \equiv P \vee (Q \vee R)$	Associative Law 2
T11	$P \rightarrow Q \equiv \sim Q \rightarrow \sim P$	Contrapositive Law

Table 8.5 *Logical equivalence for Propositional Calculus*

Once an argument form (pattern of premises and conclusion) has been validated, this form is *independent of content* (meaning) and can always be used in other contexts. Further, if the form of the premises can be matched to a subset of propositions in a context, then the conclusion can be added to the set without loss of consistency (see p. 188). The conclusion is said to *logically follow* from the premises, and the independence of the propositions (an argument remains valid *no matter what additional premises are supplied*) is the *monotonic* character of Propositional and Predicate Calculus. This monotonic feature is not always present in natural argumentation since new information (propositions) can change a conclusion. The mechanics of non-monotonic logics are not fully understood. For monotonic logic the argument form called *Modus Ponens* is:

$$(P,\ P \rightarrow Q)\ \therefore Q$$

and the argument form called *Modus Tollens* is:

$$(\sim Q,\ P \rightarrow Q)\ \therefore \sim P$$

and they are two mechanisms through which wff can be generated. Both are arrived at by inspection of the truth tables to confirm the statements:

$$(P \wedge (P \rightarrow Q)) \rightarrow Q \text{ is valid}$$

and:

$$(\sim Q \wedge (P \rightarrow Q)) \rightarrow \sim P \text{ is valid}$$

respectively (see Table 8.6 for Modus Ponens).

P	Q	$P \rightarrow Q$	$P \wedge (P \rightarrow Q)$	$(P \wedge (P \rightarrow Q)) \rightarrow Q$
T	T	T	T	T
T	F	F	F	T
F	T	T	F	T
F	F	T	F	T
T = True,		F = False		ie valid (all True)

Table 8.6 *The validity check for Modus Ponens*

In contrast, an argument is *fallacious* if it is not valid even though it may be satisfiable. An example is:

$$(Q, P \rightarrow Q) \therefore P$$

where the different interpretations are given in Table 8.7

P	Q	$P \rightarrow Q$	$Q \wedge (P \rightarrow Q)$	$(Q \wedge (P \rightarrow Q)) \rightarrow P$
T	T	T	T	T
T	F	F	F	T
F	T	T	T	F *
F	F	T	F	T
T = True,		F = False		ie A Fallacy (not valid)

Table 8.7 *A validity check for a fallacious argument*

CONSISTENCY AND *REDUCTIO AD ABSURDUM*

A set of propositions (eg the premises of an argument) is *consistent* if a model exists for their conjunction. Consistency can be checked and a model found by applying the tableau technique (Hodges, 1981) in place of the enumeration of every interpretation. The technique starts by analysing the initial statements into the component atomic propositions and connectives. One of the compound propositions is chosen, and the conjunctions (Ands) are grouped together to form a single node in a tree structure. The conjunctive nodes form a list of connected nodes in the path. Each path represents a set of formulae that must all be true if a model exists. The disjunctions (Ors) are split to form new branches.

In the process of creating this tree, as each compound

statement is selected, if there occurs a contradiction (P, ~ P) on the *same path*, then that path is *closed* since there is no interpretation that can make the path true. On completion of the tree and if all the paths are closed, then there is *no* model (the propositions are inconsistent). However, each remaining open path represents a model that satisfies the propositions. The following example (from Hodges, 1981) illustrates the technique.

P1. 'If cobalt but no nickel is present, a brown colour appears'
P2. 'Either nickel or manganese is absent'
P3. 'Cobalt is present but *only* a green colour appears'

On analysis the atomic propositions are:

Co. 'Cobalt is present'
Ni. 'Nickel is present'
Mn. 'Manganese is present'
B. 'Brown appears'
G. 'Green appears'

The compound propositions can be written as:

P1. $(Co \land \sim Ni) \rightarrow B \equiv \sim (Co \land \sim Ni) \lor B$
P2. $\sim Ni \lor \sim Mn$
P3. $Co \land (G \land \sim B)$

The tableau is shown in Figure 8.1 where the inconsistent paths are set in italics and the asterisk (*) indicates a consistent path. The equivalences of Table 8.5 are used to re-form compound propositions so that the and/or tree can be created. The consistent model is:

'Manganese is absent.
Cobalt is present and only a green colour appears.
Nickel is present.'

The tableau technique illustrated does not represent a valid argument but only a means of discovering a model for a set of propositions if it exists; it can determine consistency. The determination of a model for a set of propositions (eg see Tables 8.6 and 8.7) can be used to prove that an argument is valid. It is not sufficient just to show that an argument is

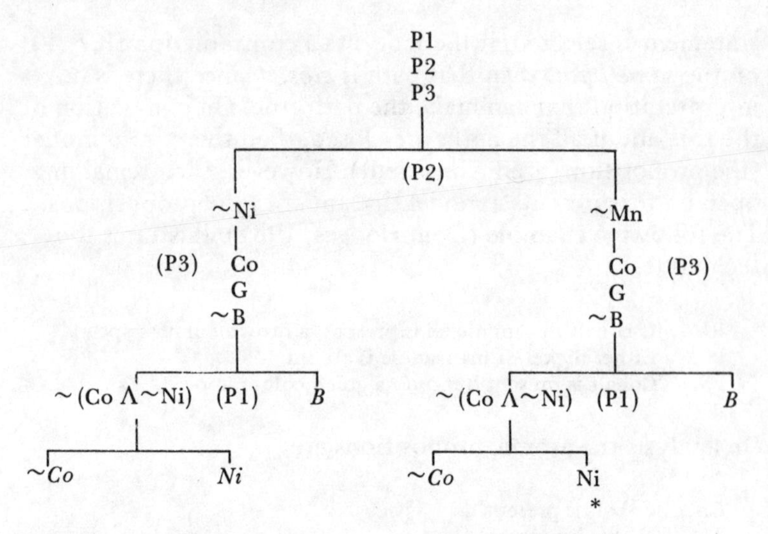

Note: $\sim(\text{Co} \wedge \sim \text{Ni}) \equiv \sim \text{Co} \vee \text{Ni}$ by T3

Figure 8.1 *The tableau of a consistent set of premises (adapted from Hodges, 1981)*

satisfiable (it may not be valid); this proof relies upon showing that if the *counterexample* has no model, then every interpretation applies to the argument and hence it is valid.

If an argument takes the form:

$$(\text{P1}, \text{P2} \ldots \text{Pn}) \therefore \text{Q}$$

then the counterexample is the conjunction of the premises and the *inverse* of the conclusion thus:

$$\text{P1} \wedge \text{P2} \wedge \ldots \wedge \text{Pn} \wedge \sim\text{Q}$$

If the argument is valid, then:

$$\text{P1} \wedge \text{P2} \wedge \ldots \wedge \text{Pn} \rightarrow \text{Q}$$

is valid, which, from T2:

$$\sim(\text{P1} \wedge \text{P2} \wedge \ldots \wedge \text{Pn}) \vee \text{Q}$$

and T3:

$$\sim\text{P1} \vee \sim\text{P2} \vee \ldots \vee \sim \text{Pn} \vee \text{Q}$$

Since every interpretation of the argument makes this proposition True, the inverse of this proposition will always be False. The inverse is:

$$\sim(\sim P1 \vee \sim P2 \vee \ldots \vee \sim Pn \vee Q)$$

which becomes from T1 and T3:

$$P1 \wedge P2 \wedge \ldots Pn \wedge \sim Q$$

and this is the counterexample. Thus, if the counterexample is shown to be inconsistent by the tableau technique, then the argument (from which the counterexample is drawn) is valid. This method of proof (that uses the counterexample) is called *reductio ad absurdum*.

It is thus possible, within the framework of Propositional Calculus, to specify a mechanistic method through which argumentation can proceed. Models can be determined, argument forms checked for validity and argument forms used to generate conclusions. All this is possible provided that the initial statements are restricted to the grammar of Propositional Calculus. However, there are many problems which require that the components of the propositions are accessed, the relationships between the components can be referenced and generalizations can be used.

FIRST-ORDER PREDICATE CALCULUS

Predicate Calculus is an extension of Propositional Calculus. It provides a richer language to express relationships between objects and the generalizations concerning these objects. Propositions can take on variables so that it is possible to express:

P	≡	'John loves Mary'
P1 (y)	≡	'John loves (someone)'
P2 (x)	≡	'(Somebody) loves Mary'
P3 (x, y)	≡	'(Somebody) loves (someone)'

where x and y can take on particular values so that :

P3 (John, Mary) ≡ 'John loves Mary'

Normally, P3 is the relationship (ie the verb) sign such that 'LOVES (John, Mary)' is used in the above example. 'LOVES' in this case is a two place atomic *predicate* (an atom), and it maps pairs of objects into the truth values True or False. A literal is either an atom or the inverse of an atom. The places in the n-tuple are *terms*, and they may be *variables*, *constants* or *functions*. Convention specifies predicates as upper case letters from the middle of the alphabet, variables as lower case letters from the end of the alphabet, constants as lower case letters from the beginning of the alphabet and functions as lower case letters from the middle of the alphabet. A function maps objects of the *universe* (where the universe is the defined set of objects under consideration) into other objects such that any particular n-tuple of the function identifies a single object. A constant is a special case of a function where $n = 0$.

Universal Quantification $(\forall x)$ provides a mechanism for ranging over the complete domain of values for x. It is equivalent to creating the conjunction (Ands) of all the propositions produced for each value of x in the domain. Thus:

$$(\forall x)\ [\text{LOVES}\ (x, \text{Mary})]$$

where x ranges over the domain of people. The statement may be translated as:

'Everybody loves Mary'

Existential Quantification $(\exists x)$ is the mechanism that indicates that at least one object satisfies the statement. This is equivalent to creating the disjunction (Ors) between the elements of the domain of people. Thus:

$$(\exists x)\ [\text{LOVES}\ (x, \text{Mary})]$$

may be translated as:

'Somebody loves Mary'

Compound expressions can be formed using the logical connectives, and the *scope* of the quantifiers can be controlled through the use of square brackets. Variables that have quantifiers are *bound* otherwise they are *free*. From the point of view of translation note that:

'All swans are white'

becomes:

$$(\forall x)\ [\text{SWAN}\ (x) \rightarrow \text{COLOUR}\ (x, \text{white})]$$

and:

'There is a black swan'

becomes:

$$(\exists x)\ [SWAN\ (x) \wedge COLOUR\ (x, black)]$$

The quantifiers are interpreted from inner to outer so that:

$$(\exists x)\ (\forall y)\ [LOVES\ (x, y)]$$

can be translated as:

'There is somebody who loves everyone'

that is, there exists an x such that every y is loved by x, and:

$$(\forall y)\ (\exists x)[LOVES\ (x, y)]$$

may be translated as:

'Everybody is loved by someone'

that is, for every y there exists an x who loves that y.

reference	logical equivalences
T12	$\sim(\exists x)[P(x)] \equiv (\forall x)[\sim P(x)]$
T13	$\sim(\forall x)[P(x)] \equiv (\exists x)[\sim P(x)]$
T14	$(\forall x)[P(x) \wedge Q(x)] \equiv (\forall x)[\ P(x)\ ] \wedge (\forall y)[Q(y)]$
T15	$(\exists x)[P(x) \vee Q(x)] \equiv (\exists x)[\ P(x)\ ] \vee (\exists y)[Q(y)]$
T16	$(\forall x)[P(x)] \equiv (\forall y)[\ P(y)\ ]$
T17	$(\exists x)[P(x)] \equiv (\exists y)[\ P(y)\ ]$

Table 8.8 *The additional equivalences for Predicate Calculus*

The tautologies of Table 8.5 still hold but require the addition of the tautologies involving quantification. These are given in Table 8.8.

Reasoning with predicates

THE INFINITY OF UNIVERSES

An interpretation of a Predicate Calculus wff is the assignment of 'values' to each constant, function symbol and predicate symbol. The values in the case of constants will be objects from the universe of discourse (the domain); for each function symbol the values will be mappings between these objects and for each evaluation of a predicate there will be an assigned True or False. However, variables are not assigned values. Since there is an infinite number of domains each of which may also be infinite then, unlike Propositional Calculus, there is an infinite number of interpretations. Thus the determination of validity and satisfiability is not simply a matter of enumeration, and it has been shown that there is *no general decision procedure to check the validity of a wff* in Predicate Calculus (Church, 1936; Turing, 1936). However, there are procedures that can, in principle, prove that a wff is unsatisfiable. If it is satisfiable, then the procedure may not terminate. Since a termination point is not always known for an unsatisfiable wff, any finite application of the procedure that has not terminated is not evidence of consistency. Predicate Calculus is semi-decidable.

There is a procedure that requires the repeated application of *sound* inference rules on the premises to generate the NIL clause. A clause is normally a disjunction of literals and contains no conjunctions. However, there is a general form of the clause which contains both conjunctions and disjunctions thus:

$$P1 \wedge P2 \wedge \ldots Pn \rightarrow Q1 \vee Q2 \vee \ldots Qm$$

where $n \geqslant 0$ and $m \geqslant 0$, Pi are joint *conditions* and Qj are alternative *conclusions* (Kowalski, 1979). This form of clause will not be referred to until rule-based systems are described on p. 204. NIL is a special clause that contains no literals (no constants, no predicates etc), and hence no interpretation can satisfy it. If the NIL clause logically follows from the premises, then the premises are unsatisfiable.

A wff Q logically follows from a set of wffs if every interpretation that satisfies that set also satisfies Q. An inference rule system or proof system is *sound* if every theorem logically follows' An inference or proof system is *complete* if it can generate all the valid formulae. Alternatively, it is *refutation complete* if it can establish inconsistency for any inconsistent set of clauses.

Universal specialization is a mechanism that allows a constant to replace a variable that is universally quantified. Informally, this simply states that if every object has a property (relationship), then any of the objects can be chosen to act as a representative for the complete set. What is true for that single object must be true for the whole domain. Thus, where A is a term then:

$$((\forall x)\ [P(x)],\ A)\ \therefore\ P(A)$$

This combined with Modus Ponens gives a complete and sound inference system. Hence:

$$((\forall x)\,[P(x) \rightarrow Q(x)],\ P(A)) \therefore Q(A)$$

It is possible through the universal specialization rule and other mechanisms to produce a set of quantifier-free clauses. Such a set of clauses is the basic requirement for the important general inference mechanism *resolution.*

CLAUSE FORMATION

Clause formation is the creation of clauses from any wff. The procedure follows nine distinct steps that must be done in order, when applicable. Usually, only a few steps are applicable. The steps are:

1. *Eliminate implication symbols:* all implication symbols are removed with the aid of T2. ($\sim P \vee Q \equiv P \rightarrow Q$).
2. *Reduce scope of the negation symbol:* thus the negation sign is moved inside brackets next to each atomic formula to which it applies. T1,T3, T4, T12 and T13 are used.
3. *Standardize the variables:* this is the process of renaming every variable so that each quantifier has its own. This ensures that the variables are linked to their quantifiers independently of the form of the wff. T16 and T17 are used.
4. *Eliminate the existential quantifiers:* this requires the use of the Skolem functions. A Skolem function is an additional mapping function that replaces a variable bound to the existential quantifier. For example:

   $$(\forall x)\,(\exists y)\,[LOVES\,(x, y)]$$

   Now a function g can be defined so that $y = g(x)$ and every x will automatically map into some y which exists according to the formula (chooses a y). For example, we have:

   $$(\forall x)\,[LOVES\,(x, g\,(x))]$$

   and when x = John then g (John) may map to Mary (ie take on the value Mary). However, if the quantifiers are the other way around:

   $$(\exists y)\,(\forall x)\,[LOVES\,(x, y)]$$

   then a *Skolem constant* is used since there is only one individual (A) referenced. Thus, we have:

   $$(\forall x)\,[LOVES\,(x, A)]$$

   This rule concerning the relative scope of the quantifiers can be extended from two to many. Skolem functions are needed for each universally quantified variable whose scope includes an existentially quantified variable, otherwise a Skolem constant is used. It can be shown that proofs involving Skolem functions are generally applicable.

5. *Convert to Prenex Form:* this is simply the movement of all universal quantifiers outwards to the front of the wff. Since every quantifier has its own variable, the correct scope is not lost.
6. *Put the main body of the wff into conjunctive normal form:* thus, a set of conjoined clauses is produced. T6 will be used.
7. *Eliminate the universal quantifiers:* this is simply the process of removing them from the front of the wff. Convention assumes that all unbound variables are universally quantified, so there is no need to explicitly state this.
8. *Eliminate all the* $\wedge$ *symbols:* each clause is now separated. Since interpretations are always concerned with the conjunction of the premises, clauses may be separated. This is just a notational convenience.
9. *Standardize the variables apart:* this is the process of renaming every variable so that no variable occurs in more than one clause.

The procedure converts the clauses so that a refutation of these clauses can also be a refutation of the original formulae. The clauses are now ready for the application of the resolution inference procedure, and illustrations of these steps will be given on p. 199. However, as for the tableau technique, resolution depends upon the comparison of pairs of clauses. Unfortunately, comparisons are not obvious, since the variables may take on many forms. Thus, P(x, z) is equivalent to P(y, g(y)) provided that certain substitutions are made. The procedure for making correct substitutions is called *unification.*

PATTERN MATCHING

Unification is concerned with pattern matching. It is a method through which two symbol strings (a wff is a symbol string) can be shown to be similar. However, there is the restriction that some symbols must be matched unchanged (ie predicates and constants), whereas other symbols may be substituted, according to certain rules. Ultimately, two patterns must be made exact images if there is to be a mechanical matching procedure. Unification provides the lawful transformations.

A *substitution instance* is a set of term/variable pairs.Each pair indicates the variable to be substituted and the term that replaces it. The terms may be constants, functions or variables. Usually a substitution is not allowed to contain two or more substitutions for one variable. Thus, (-- t1/x . . . t2/x . . .) is not allowed, and also x/x is not allowed. For example, apply the substitution s1:

s1 = (k/x, w/y, f(w)/u, B/z)

(where k/x may be spoken of as 'put k wherever x') in the expression:

P[x, g (y)] V Q [z, u, y]

and it becomes:

P [k, g (w)] V Q [B, f (w), w]

Two substitutions, s1 and s2, may be amalgamated by applying s2 to s1 and adding all term/variable pairs not used to modify s1. For example:

(g (x, y)/z) (A/x, B/y, C/w, D/z)

becomes:

(g (A, B)/z, A/x, B/y, c/w)

The associative law but not the commutative law applies to substitutions. Hence:

(E.s1).s2 = E. (s1.s2)

and:

(s1.s2).s3 = s1. (s2.s3)

but:

s1.s2 is generally NOT the same as s2.s1.

Given two patterns (compound predicates, say), unification will find a substitution (if it exists) that when applied to both patterns will make them equal. A unification pattern matcher will determine a consistent set of bindings for the variables in both patterns that results in identical patterns. The substitution that achieves a pattern match is called a *unifier* and the 'simplest' substitution is called the *most general unifier* (mgu). The simplest substitution is where any other unifier is a special case, and the special case can always be reduced to the mgu. (A LISP function for determining the mgu is given in Wilensky, 1984.)

Unification and information retrieval (eg Flexible Language Interpreter or FLIN) are related in that unification matches two patterns and returns the bindings (substitutions), whereas information retrieval scans a set of patterns with a single pattern (selector) and returns a set of bindings (the retriever). However, unification allows the case for both patterns to have variables and constants, whereas information retrieval restricts the terms to variables in the selecting pattern and constants in the scanned set. Information retrieval has the additional power to define a pattern class by using Boolean matching in the selector. For example, if a Pseudo Retrieval Language (PRL) is defined as a single statement of the form:

GETB	⟨list of n-tuples⟩ (Get the bindings for these patterns)
DISPLAYB	⟨list of variable sets⟩ (Display the bindings in these orders)
FOR	⟨boolean⟩ (Apply this selector after the pattern match)
IN	⟨list of n-tuples⟩ (List of patterns for matching)

where attributes are referred to by their position in an n-tuple, then the different methods can be compared. Each n-tuple is treated independently in the ⟨list of tuples⟩. The ⟨list of variable sets⟩ refers to the variable bindings to be returned. Where there are no bindings the variable name is returned. Each variable set gives an alternative returning pattern.

1. *Unify* compares (1, x, 3, y) with (u, 2, 3, 4) say and returns (1/u, 2/x, 4/y). This would be expressed in the PRL as:

   ```
   GETB '((1, x, 3, y))'
   DISPLAYB '((u, x, y))'
   FOR T
   IN '((u, 2, 3, 4))';
   ```
   with the response:
   ```
   ((1, 2, 4)).
   ```

2. *Information retrieval* allows a selector, thus:

   ```
   GETB '((1, x, z, y))'
   DISPLAYB '((x, y))'
   FOR 2 < z < 5
   IN '((1, 2, 3, 4) (2, 4, 6, 8) (1, 3, 4, 2) (1, 2, 6, 3))';
   ```
   and returns:
   ```
   ((2, 4) (3, 2)).
   ```

3. *Relational operations* may also be expressed in this form so that the natural join between two binary relations (expanded to 4-tuples for this purpose) becomes:

```
GETB '((a, b, 1, 3) (a, b, 2, 3))'
DISPLAYB '((a, b, y))'
FOR b = x
IN '((7, 1, x, y) (6, 1, x, y) (3, 2, x, y))';
```

and this will give:

```
((7, 1, 3) (6, 1, 3) (3, 2, 3))
```

The above statement also includes projection. Division can also be expressed if PRL is allowed to make links between tuples in a relation as in:

```
GETB '((b1, 3) (b2, 4))'
DISPLAYB '((b1) (b2))'
FOR b1 = b2
IN '((4, 3) (4, 2) (3, 5) (4, 4))';
```

and this is:

```
((4)).
```

RESOLUTION REFUTATION

Resolution is an inference method that is applied to pairs of clauses. The method is complete. To apply resolution a wff must first be converted into clauses. If one of the clauses, through the application of unification, contains the inverse of a component of another, then the inverses are equated. The result is a clause with the component and its inverse removed and the disjunction of all remaining literals. If there are no inverses, then there can be no reduction in the number of clauses (ie no resolution). For example:

$$(\forall y)\ (\exists x)\ [P(x) \wedge P(y) \rightarrow Q(y) \vee R(y)]$$

forms the clauses:

$$\sim P(x) \vee Q(x)$$
$$P(y) \vee R(y)$$

When these are unified the mgu is:

$$(y/x)$$

The clauses become:

$$\sim P(y) \vee Q(y)$$
$$P(y) \vee R(y)$$

and the components that cancel (ie because of Modus Ponens) are $\sim P(y)$ and $P(y)$, leaving:

$$Q(y) \lor R(y)$$

which is called the *resolvent*.

Resolution refutation is the use of *reductio ad absurdum* with the resolution inference method. Similarly to that of the tableau technique, a counterexample is produced by inverting the conclusion and conjoining this to the premises. However, instead of producing a tableau with all paths closed, resolution is applied repeatedly to *pairs* of clauses. Each new clause is added to a growing set of clauses for possible selection until the NIL clause is generated. When this occurs, it is known that the counterexample is unsatisfiable, and therefore the original statements are valid.

Consider the following examples:

'There is someone who's going to pay for all the services. *Therefore* each of the services is going to be paid for by someone.'

The question is 'Is the above statement valid?' First, the statements are put into Predicate Calculus form, thus:

$$P(x) \equiv \text{'x is a person'}$$
$$S(x) \equiv \text{'x is a service'}$$
$$G(x, y) \equiv \text{'x is going to pay for y'}$$

It must be shown that if the premise is:

$$(\exists x) \left[P(x) \land (\forall y) [S(y) \rightarrow G(x, y)]\right]$$

then the conclusion is:

$$(\forall y) \left[S(y) \rightarrow (\exists x) [P(x) \land G(x, y)]\right]$$

This may be done by showing that the following formula is valid:

$$\left[(\exists x) \left[P(x) \land (\forall y) [S(y) \rightarrow G(x, y)]\right]\right] \rightarrow (\forall y) \left[S(y) \rightarrow (\exists x) [P(x) \land G(x, y)]\right]$$

A counterexample is produced by conjoining the inverted conclusion with the premise:

$$(\exists x) \left[P(x) \land (\forall y) \left[S(y) \rightarrow G(x, y)\right]\right] \land \sim (\forall y) \left[S(y) \rightarrow (\exists x) \left[P(x) \land G(x, y)\right]\right]$$

The expression is now converted to clauses as specified on p. 195.

1. Eliminate implication symbol.
   $(\exists x)\left[P(x) \wedge (\forall y)\left[\sim S(y) \vee G(x,y)\right]\right]$
   $\wedge \sim(\forall y)\left[\sim S(y) \vee (\exists x)\left[P(x) \wedge G(x,y)\right]\right]$
2. Reduce scope of negation symbol.
   $(\exists x)\left[P(x) \wedge (\forall y)\left[\sim S(y) \vee G(x,y)\right]\right]$
   $\wedge (\exists y)\left[S(y) \wedge (\forall x)\left[\sim P(x) \vee \sim G(x,y)\right]\right]$
3. Standardize variables.
   $(\exists x1)\left[P(x1) \wedge (\forall x2)\left[\sim S(x2) \vee G(x1, x2)\right]\right]$
   $\wedge (\exists x3)\left[S(x3) \wedge (\forall x4)\left[\sim P(x4) \vee \sim G(x4, x3)\right]\right]$
4. Eliminate the existential quantifiers.
   $P(A1) \wedge (\forall x2)\left[\sim S(x2) \vee G(A1, x2)\right]$
   $\wedge \left[S(A2) \wedge (\forall x4)\left[\sim P(x4) \vee \sim G(x4, A2)\right]\right]$
   where A1 and A2 are Skolem constants.
5. Convert to Prenex Standard Form.
   $(\forall x2)(\forall x4)\left[P(A1) \wedge (\sim S(x2) \vee G(A1, x2))\right.$
   $\left.\wedge S(A2) \wedge (\sim P(x4) \vee \sim G(x4, A2))\right]$
6. Put the main body of the wff into conjunctive form (no action required since it is already in that form).
7. Eliminate the universal quantifiers.
   $P(A1) \wedge (\sim S(x2) \vee G(A1, x2)) \wedge S(A2) \wedge (\sim P(x4) \vee \sim G(x4, A2))$
8. Eliminate all the $\wedge$ symbols. The 'goal' clauses are set in italics. Strictly, the premises and negated conclusion should have been dealt with separately when converting to clauses if the goal clauses are to be singled out.
   (P(A1), (~S(x2) V G(A1, x2)), *S(A2), (~P(x4) V ~G(x4, A2)))*
9. Standardize the variables apart (no action since no variable is found in more than one clause).

Resolution can now be applied to pairs of clauses drawn from the 'database' of clauses. Each new clause is added to the database. However, to make clear the stages in the compound argument a refutation tree is drawn that records the clause pairs chosen, the mgu and the new clause produced. The root of the tree should be the NIL clause if the proof is successful. Figure 8.2 shows the refutation tree.

The resolution refutation procedure has generated a set of consistent substitutions as a side effect of the unification. These substitutions can be recovered if the conclusion is turned into a question. This would be:

'There is someone who's going to pay for all the services.
Therefore each of the services is going to be paid for by someone.
Who is this someone?'

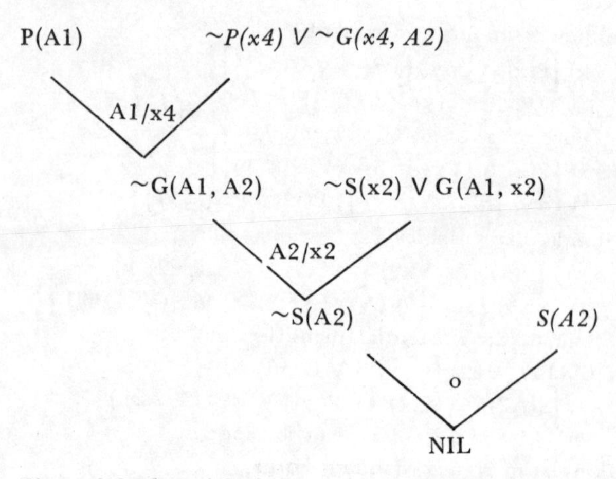

Figure 8.2 *The refutation tree for services example*

To recover the substitutions, the refutation tree has the negation of each of its goal clauses appended to those clauses. These clauses have been underlined in Figure 8.2. The *answer tree* is illustrated in Figure 8.3. The substitutions found in the refutation tree are then applied to the extended clauses.

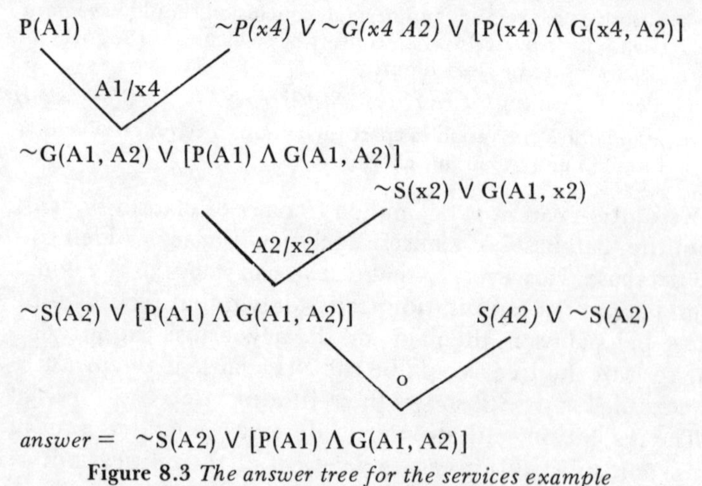

answer = ~S(A2) V [P(A1) Λ G(A1, A2)]

Figure 8.3 *The answer tree for the services example*

The answer can be converted (using T2) to:

$$S(A2) \rightarrow (P(A1) \land G(A1, A2))$$

and in English this will be:

> 'The services are paid for by the person who pays for one of the services'

which is the best answer obtainable given the information in the example. More interesting answers are produced given more complex cases.

CONTROL STRATEGIES

Resolution refutation falls within the paradigm of the production rule system (see Chapter 7) where the database of clauses is transferred into a large database of clauses through the application of resolution. The choice of transformation is indicated by the pair of clauses selected to be resolved, and the goal is a database that contains the NIL clause. The system is commutative and uninformed so there is no need for backtracking or heuristics. However, there are strategies that will reduce the search for the NIL clause.

The *breadth-first strategy* is where every pair of clauses in the initial database (the base set) is used to generate a set of resolvents. These resolvents are then added to the database, and every pair of clauses in that database (except pairings that have already been done) are then resolved thus generating a further set of resolvents. These resolvents are then added to the database and so on. This strategy is complete but very inefficient.

The *set-of-support strategy* is similar to the breadth-first strategy except that the pairs of clauses are limited so that one of the pairs is always a clause of the negated conclusion or a descendent of the negated conclusion. The set of clauses related to the negated conclusion is called the set of support. This strategy is complete and may be viewed as 'backward' reasoning because it starts with the conclusion. However, the 'goal' of the system is the NIL clause.

Unit-preference strategy is an efficiency modification to the set-of-support strategy in that a single-literal clause (the unit) is chosen to be one of the pair in a resolution. This is because the resolvents will have fewer literals than the parents, and hence it is 'nearer' to the NIL clause, and this ordering of the clause pairings will generally achieve the goal sooner.

Linear-input form strategy resolves pairs of clauses where one of the pair is always from the base set. This is not complete since there are some unsatisfiable sets of clauses that do not produce the NIL clause under this regime. However, it is simple and efficient.

Ancestry-filtered form is a modification of the linear-input form by allowing the option of the pairings of two clauses that share the same ancestor. This retains the efficiency of linear-input form, and the extra freedom of pair choice makes it complete. The sharing of a common ancestor for pairs of clauses during resolution encourages the occurrence of merging two literals into one. This takes the system nearer the NIL clause goal as for the unit-preference strategy.

Strategies may be combined, and, in particular, the set-of-support strategy can be linked with linear-input or ancestry-filtered form. Informed variations of these strategies may be used through knowledge of the 'meaning' of certain predicates such that their truth value can be pre-assigned. Clauses that subsume other clauses may also be eliminated from consideration.

Rule-based systems

A SIMPLE SCHEME

The dominant scheme for encoding 'expert knowledge' is the rule. A rule is a clause (see p. 194) that contains an implication and at most a single conclusion. Such clauses are called Horn Clauses and any problem in logic can be re-expressed in the Horn Clause form (Kowalski, 1979). The literals to the left of the implication sign are called the *antecedent*, and the literals to the right are called the *conclusion*. Sometimes the conclusion is an *action* or a *suggestion*. Frequently, the rules are expressed as:

IF antecedent *THEN* conclusion

and an example from a medical knowledge base (MYCIN) is:

IF stain is gram positive
AND morphology is coccus
AND growth is chains
THEN suggestive evidence that organism is streptococcus

There is a simple procedural view of reasoning with rules that illustrates the essential nature of such systems. This may best be described by a trivial example. Consider the following two rules:

R1. *IF* patient P exhibits symptoms X *AND* Y

THEN patient P suffers from disease type A
R2. *IF* patient P suffers from disease type A
AND patient P exhibits symptom Z
THEN patient P suffers from disease type B

The system can determine if patient P suffers from disease type B by either *forward* or *backward* chaining through the rules. Forward chaining starts with the antecedents of a rule (R1 say) and attempts to establish if R1 is 'True'. Thus, if X and Y are 'True', then conclusion:

$$X \wedge Y \rightarrow A$$

is recorded. Rules that have A in their antecedent are examined next (R2), and the truths of the associated predicates (Z) are determined. If these are also 'True', then the information is 'attached' to the previous conclusion and the final result is given. Thus:

R1. $X \wedge Y \rightarrow A$

$\updownarrow$

R2. $A \wedge Z \rightarrow B$

$\therefore B$

Backward chaining, on the other hand, starts with a proposed conclusion (B say) and finds a rule that has just this conclusion (R2). The antecedent is established by first finding rules that have elements of the antecedent as conclusions (R1) and second by determing the truth values (usually by 'asking') of the remaining literals (Z).

This simple procedural view hides many important issues. In particular, the link between rule-based systems and the formal analysis derived from Predicate Calculus. If this link is not made correctly and is not understood, then there is a danger that the design of a knowledge-based system will have no basis for accreditation (and as such cannot be relied upon). Further, there will be no foundation for a clear development programme, and many subtle uses of the 'rule' will be lost.

THE NATURE OF RULES

The logical context of a formula is independent of its form. The form of a formula may be used to encode a preferred mode of application. In particular, the Horn Clause form (rule form) is an easy way of representing knowledge and contains a strong suggestion as to how it may be employed during inference. This is particularly important since the resolution refutation system has no control information associated with clauses (other than content-free strategies). However, there are the assumptions in this view that the control information provided by the rule form is appropriate to both the method of inference used within a program and the objectives which the rule is meant to help achieve. Neither of these assumptions has been formally confirmed.

The improved performance of a rule-based system over the resolution refutation system comes from the way in which a rule-based system operates. Rules do not interact freely, as do clauses, since they cannot make inferences directly. Rules can only make inferences by a one at a time application to a highly structured database. This single application of a rule to an evolving and increasingly complex statement (the database) restricts the possibilities. Hence the search space is much reduced. Further, the clausal form tends to be highly redundant in that a single step for a rule may represent many steps in resolution.

Figure 1.1 (see p. 21) shows the different types of asserted knowledge. Rules (hypotheses) contain implicitly structural and/or reasoning (transformational) knowledge. In the simple example given on p. 204, there is the implicit structural relationship between disease types A and B in that B is a subcategory of A. The reasoning and control knowledge is also implicit in that inference (behaviour) is occurring within certain constraints given by the form of the rule. Facts contain constants and variables but they typically represent specific cases (ie facts are like propositions).

The task of a rule-based system is to 'prove' that a goal wff (a goal hypothesis) is consistent with (observed) facts and (given) rules. The different types of knowledge are illustrated in Figure 8.4 (cf Figure 1.1). Note that there is implicit structural and control knowledge contained in the transformational rule set and the explicit taxonomic and categorical knowledge is provided by the structural rule set.

THE METHODS FOR RULE-BASED SYSTEMS

Forward and backward systems use slightly different methods, and each system is used to achieve different kinds of goals. Forward systems are specifically designed to determine a *disjunction* of literals, such as:

R2.2/1 ~BULLDOG (Butch) V FRIENDLY(Butch) V NOISY(Butch)

and the backward systems are used to determine a *conjunction* of literals, such as:

R2.1 CAT(Tom) Λ ~AFRAID(Tom, Butch) Λ DOG(Butch)

1. *FACTS* (observed): given synthetic 'facts'	
R1.1	Tom meows
R1.2	If Butch is a dog then either he does not bark and wags his tail or he barks and bites
R1.3	Butch is a dog
2. *FACTS* (derived): 'goal' statements that are consistent	
R2.1	The cat Tom is not afraid of the dog Butch
R2.2	If Butch is a bulldog then he is either friendly or noisy
3. *INFERENTIAL RULES* (transformational)	
R3.1	If something wags its tail and is a dog then it is friendly
R3.2	If something is friendly and does not bark then everybody is not afraid of it
4. *INFERENTIAL RULES* (structural)	
R4.1	Dogs and cats are animals
R4.2	Cats meow
R4.3	Anything that barks is noisy
R4.4	All bulldogs are dogs

Figure 8.4 *Examples of different kinds of asserted knowledge*

The general procedure, in both cases, is to prepare the facts, goal and rules *specifically* for the system to be used. The starting point ('facts' for the forward system and 'goal' for the backward system) is an AND/OR graph whose rules of construction are different for each system. The graph is then extended by the addition of the rules that are linked according to a common (to both systems) unification technique. There is no restriction on the number of times a rule may be used. At anytime, different conclusions may be 'read off' the structure but the objective is to achieve a reading that will match the 'goal' for the forward system or the 'facts' for the backward system. The objectives can only be said to have been met if the unification consistency check (see p. 212 for a description of unifying composition) shows a coherent result.

PREPARATION OF FACTS FOR THE FORWARD SYSTEM

The facts are formed into a structural AND/OR graph global database by first converting them into a single wff that contains no quantifiers and no implications. This procedure is a subset of the scheme used for resolution refutation, and it is usually needed if the facts contain variables or implications. For example, it may be a fact that: 'There is an animal to which the dog Butch is never friendly' and this may be expressed as:

$$(\exists x)\ [\mathrm{ANIMAL}(x) \wedge \mathrm{DOG}(\mathrm{Butch}) \wedge \sim \mathrm{FRIENDLY}(\mathrm{Butch})]$$

or R1.2 and:

'Every animal that moves is chased by Butch'

where the latter can be expressed as:

$$(\forall x)\ [\mathrm{MOVES}(x) \wedge \mathrm{ANIMAL}(x) \rightarrow \mathrm{CHASES}(\mathrm{Butch}, x)]$$

'Facts' of this kind may also be considered as context-dependent hypotheses. The fact conversion scheme for the forward system is:

1. Eliminate implication symbols.
2. Reduce the scope of negation.
3. Standardize the universally quantified variables by renaming.
4. Eliminate the existentially quantified variables by Skolem functions.
5. Drop the universal quantifiers.
6. Standardize variables apart in the main conjunctions only.

Thus:

$$Q(u, A) \wedge \ [\![R(u) \wedge P(u)] \vee S(A, u)]\!]$$

may be standardized apart to some extent by:

$$Q(x, A) \wedge \ [\![R(y) \wedge P(y)] \vee (S(A, y)]\!]$$

The subexpression $[R(y) \wedge P(y)]$ must be linked with the literal $S(A, y)$ because of the disjunction. The reason for this is the manner in which consistency is maintained.

The example facts given in Figure 8.4 can now be prepared for the conversion into a structural global database:

$$\begin{aligned}&\mathrm{DOG}(\mathrm{Butch}) \wedge \mathrm{MEOWS}(\mathrm{Tom})\\ &\quad \wedge [\mathrm{DOG}(\mathrm{Butch}) \rightarrow [\![[\sim \mathrm{BARKS}(\mathrm{Butch}) \wedge \mathrm{WAGS\text{-}TAIL}(\mathrm{Butch})]\\ &\qquad\qquad\qquad\qquad\quad \vee [\ \mathrm{BARKS}(\mathrm{Butch}) \wedge \mathrm{BITES}(\mathrm{Butch})]]\!]]\end{aligned}$$

and this becomes:

DOG(Butch) Λ MEOWS(Tom)
Λ [~DOG(Butch)
V [~ BARKS(Butch) Λ WAGS-TAIL(Butch)]
V[BARKS(Butch) Λ BITES(Butch)]]

The AND/OR graph is created from this wff by expressing it in terms of k-connectors and 1-connectors. However, there is an inversion of the usual interpretation of these connectors because they refer to the consistency checking rather than to the 'meaning' of the logical connectives. Thus:

1. Disjuncts (V) are linked together with k-connectors.
2. Conjuncts (Λ) are each 1-connectors.
3. Leaf nodes will be literals.

Figure 8.5 shows the facts expressed as an AND/OR graph.

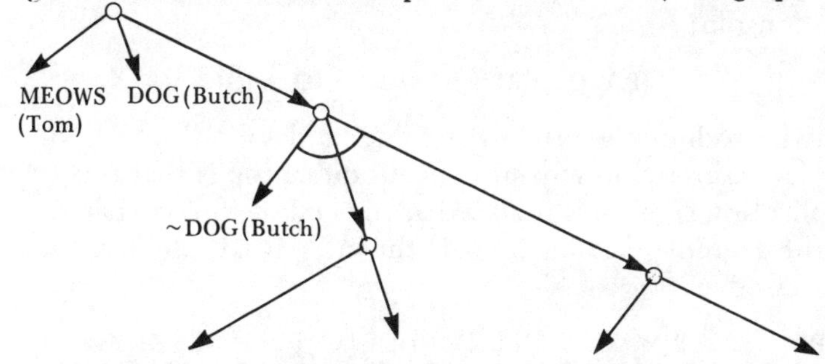

Figure 8.5 *The facts from Figure 8.4 given in AND/OR graph form*

The graph may be 'read' by examining a path (as defined for AND/OR graphs). Each path represents a possible solution (a consistent wff), and the set of paths is the set of solution graphs. In this example the solution graphs are:

DOG(Butch)
MEOWS(Tom)
~DOG(Butch) V ~BARKS(Butch) V BARKS(Butch)
~DOG(Butch) V WAGS-TAIL(Butch) V BARKS(Butch)
~DOG(Butch) V ~BARKS(Butch) V BITES(Butch)
~DOG(Butch) V WAGS-TAIL(Butch) V BITES(Butch)

and these are the clauses that would have resulted in the normal clause formation process for resolution. The AND/OR graph is a non-redundant representation. However, if the structure contains *variables*, then this representation is not general because the variables are not all standardized apart. The degrees of freedom for possible solutions are thus restricted since a variable is now tied to having the same value in groups of statements where it would otherwise have had the potential for different values. This makes the rule-based systems *not complete*.

FORWARD RULES

The AND/OR graph that forms the global database is transformed by applying a rule. The rules, however, must be in the right form in order to append themselves to the leaf nodes. Since these nodes are literals, the rules must be of the form:

$$\text{Literal} \rightarrow \text{wff}$$

where the wff is an AND/OR subgraph.

Any rule can be converted into this form by using the relationship:

$$(P \vee Q \rightarrow R) \equiv ((P \rightarrow R) \wedge (Q \rightarrow R))$$

(which produces two rules from one) and a modification of the fact conversion scheme. The modification is that the implication symbol is replaced at the end of the six stages. In the example (see Figure 8.4) the inferential rules may be expressed as:

R3.1 (∀x1)[[WAGS-TAIL(x1) ∧ DOG(x1)] → FRIENDLY(x1)]
R3.2 (∀x2) (∀x3) [[FRIENDLY(x2) ∧ ~BARK(x2)] → ~AFRAID(x3, x2)]
R4.1 (∀x4) [[DOG(x4) ∨ CAT(x4)] → ANIMAL(x4)]
R4.2 (∀x5) [MEOWS(x5) → CAT(x5)]
R4.3 (∀x6) [BARKS(x6) → NOISY(x6)]
R4.4 (∀x7) [BULLDOG(x7) → DOG(x7)]

and these become:

R3.1/1 WAGS-TAIL(x1) → [FRIENDLY(x1) ∨ ~DOG(x1)]
R3.2/1 ~BARKS(x2) → [~FRIENDLY(x2) ∨ ~AFRAID(x3, x2)]
R4.1/1 DOG(x4) → ANIMAL(x4)
R4.1/2 CAT(x5) → ANIMAL(x5)
R4.2/1 MEOWS(x6) → CAT(x6)
R4.3/1 BARKS(x7) → NOISY (x7)
R4.4/1 BULLDOG(x8) → DOG(x8) ≡ ~DOG(x8) → ~BULLDOG(x8)

Rule R4.4/1 will be given in its contrapositive form (as can any of the rules provided there is a single literal to the left of the implication sign) because there is some ‘knowledge’ of how this rule will be used.

ACHIEVING GOALS

Achieving a goal wff in a direct system of proof is like determining the answer wff from the answer tree in resolution refutation. However, instead of taking two steps (refute by finding NIL then add the negation of the original clause etc) only the one step is taken. The desired clause is found directly through valid argument.

The goals are thus formed differently to facts in that *universal* quantifiers are treated as though they were existential quantifiers and vice versa. This is related to the original refutation technique that inverts quantification. Thus, universal quantified variables are replaced by Skolem functions, and existential quantifiers are renamed where possible (between disjuncts as for T15). The goal statements of Figure 8.4 before analysis were:

R2.1/g 'Is there a cat who is not afraid of a dog?'

and:

R2.2/g 'Is there a bulldog which is either friendly or noisy?'

These would be represented as:

R.2.1/g1 $(\exists x9)(\exists x10)[CAT(x9) \wedge DOG(x10) \wedge \sim AFRAID(x9, x10)]$
R2.2/g2 $(\exists x11)[BULLDOG(x11)$
$\rightarrow [FRIENDLY(x11) \vee NOISY(x11)]]$

and these will be converted to wffs that are a disjunction of literals. However, only R2.2/g2 can be converted for the forward system, and this becomes:

R2.2/g2/1 $\sim BULLDOG(x11) \vee FRIENDLY(x11) \vee NOISY(x11)$

Selected rules can now be applied to the AND/OR graph by matching the single literals on the left side of the implication to that of the leaf nodes of the growing graph. The matching is achieved through unification, and each successful unification is marked by a special arc (the broad arrow in the AND/OR graph). Once matched, the rule is added to the graph as an extra substructure. The goal is achieved when there is a consistent 'reading' of the graph that matches the form of the converted goal. Figure 8.6 illustrates the result of the forward system on the above example.

The goal literals are attachable to the penultimate graph in a manner that would generate a clause of the right form. The important issue is that the unifications used at each stage

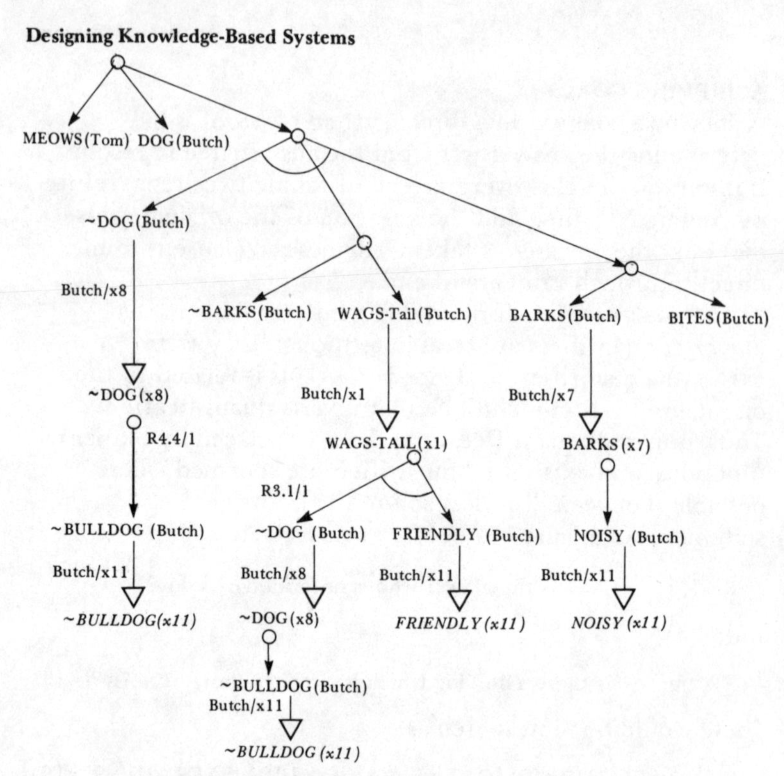

Figure 8.6 *The achieved goal in the forward system*

are consistent. The consistency check is the last step (if successful) in the forward system.

UNIFYING COMPOSITION

The procedure for checking that a collection of unifications for a particular solution is coherent also requires the process of unification. The collection to be considered in the AND/OR graphs of either forward or backward systems is that collection of unifications which lies on the same path. That is the path defined for such a graph. In the example, the solution path includes all the underlined leaf rules in Figure 8.6 and the set of unifications involved is:

((Butch/x8), (Butch/x11), (Butch/x1), (Butch/x8), (Butch/x11), (Butch/x7), (Butch/x11))

This can be reduced to the set:

((Butch/x8), (Butch/x1), (Butch/x7), (Butch/x11))

The procedure for checking consistency is to make two lists. One of all the terms, and the other of all the variables. These will be:

(Butch, Butch, Butch, Butch)

and:

(x8, x1, x7, x11)

The unification is consistent if there is a unifier for these two sets, and in this case the unifer is simply:

(Butch/x8, Butch/x1, Butch/x7, Butch/x11)

This unifier is the mgu applied to the solution taken from Figure 8.6:

~BULLDOG(Butch) V FRIENDLY(Butch) V NOISY(Butch)

which translates into the derived fact R2.2.

This particular example is relatively easy, but the procedure works for more complex and less obvious examples. Thus, if the unifiers were:

((A/x), (B/x))

then the two sets are:

(A, B), (x, x)

and there is *no* mgu for these two sets. Consequently, the unifications are inconsistent. On the other hand, consider the two unifiers:

((f(g(x1))/x3, f(x2)/x4), (x4/x3, g(x1)/x2))

As before, the procedure is to create two sets of terms and variables ignoring the structure of the set:

(f(g(x1)), f(x2), x4, g(x1))
(x3, x4, x3, x2)

These are unified by the substitutions:

(f(g(x1))/x3, f(g(x1))/x4, g(x1)/x2)

as can be tested by applying them in the given order.

This method of unifying composition is applicable to both forward and backward rule-based systems. The operator is both associative and commutative since the order in which the unification takes place in the development of the global database does not matter. If a rule is applied more than once, then the variable should be renamed. Otherwise the result is overconstrained.

Consistency may be regained by backtracking to a point where the unification substitutions are no longer inconsistent. Thus, in the inconsistent example given, the system could backtrack to either before A/x or B/x and could then try applying different rules. For example, we may have the 'fact' that there is only one thing that is either a cat or a dog thus:

$$CAT(x) \vee DOG(x)$$

and the two 'rules':

R1. $CAT(Tom) \rightarrow ANIMAL(Tom)$
R2. $DOG(Butch) \rightarrow ANIMAL(Butch)$

The graph of this example is shown in Figure 8.7. This deliberately overconstrained example will only give the solutions:

$$ANIMAL(Tom) \vee DOG(Tom)$$

and:

$$ANIMAL(Butch) \vee CAT(Butch)$$

since the constraints are that the same constant must appear in both literals and that both the solutions must be true, this is the only way that it can be done.

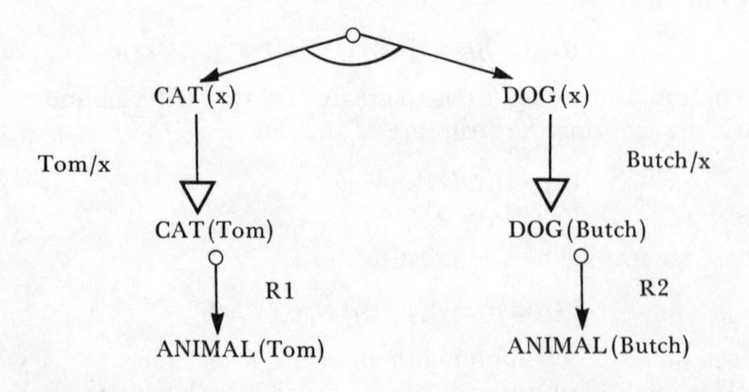

Figure 8.7 *Inconsistent substitutions (adapted from Nilsson, 1980)*

BACKWARD RULES

There is a symmetry in the resolution refutation system in that goals can be treated as assertions (facts and rules) and assertions can be treated as goals. In the normal approach, goals are negated and added to assertions where the resolution is applied until the NIL clause is found. The empty clause is interpreted as False, and the set of clauses is considered unsatisfiable. The alternative approach is to negate the assertions and add them to the goal. In this case, resolution is still applied until the NIL clause is found, as before, but this clause is now considered as always True. Validity (always true) is the converse of unsatisfiability (always False), and the dual resolution refutation system states that every interpretation that is unsatisfiable with respect to the assertions should also be unsatisfiable with respect to the goal. If there is no interpretation that cannot be satisfied, then the goal is 'refuted' in that it will always be true for all the models of the original (unnegated) assertions.

Backward rules take advantage of this symmetry, and the system starts with the goal as the global database. The goal, as for the forward system, has the universally quantified variables standardized apart (with the relations as before). The major differences from that of the forward system are:

1. Disjunctions are 1-connectors.
2. Conjunctions are k-connectors.
3. Backward rules are made in the form:

   $$\text{wff} \rightarrow \text{literal}$$

   and employ the tautology:

   $$(P \rightarrow R \wedge Q) \equiv ((P \rightarrow R) \wedge (P \rightarrow Q))$$

   therefore two rules can be produced from the single rule.
4. The solutions (hypergraph paths) are conjunctions of literals.

The procedure for applying the rules is similar to the forward technique except that the AND/OR graph terminates in fact literals. Unifying composition checks consistency as before.

The example (see Figure 8.4) can be adjusted to a backward system by modifying the rules so that the rules have a single literal on the right. The rules concerned are:

R3.1/1 WAGS-TAIL(x1) Λ DOG(x1) → FRIENDLY (x1)
R3.2/1 ~BARKS(x2) Λ FRIENDLY(x2) → ~AFRAID(x3, x2)

The facts also need changing in the same way, and R1.2/1 in particular can be modified:

DOG(Butch) Λ ~[BARKS(Butch) Λ BITES(Butch)] →
[~BARKS(Butch) Λ WAGS-TAIL(Butch)]

This becomes facts with single literals on the right:

R1.2/2 DOG(Butch) Λ ~BARKS(Butch) → ~BARKS(Butch)
R1.2/3 DOG(Butch) Λ ~BITES(Butch) → ~BARKS(Butch)
R1.2/4 DOG(Butch) Λ ~BARKS(Butch) → WAGS-TAIL(Butch)
R1.2/5 DOG(Butch) Λ ~BITES(Butch) → WAGS-TAIL(Butch)

A goal that is achievable in the forward system is R2.1/g1, and this is expressed as:

CAT(x9) Λ DOG(x10) Λ ~AFRAID(x9, x10)

which is converted into the initial global database illustrated in Figure 8.8.

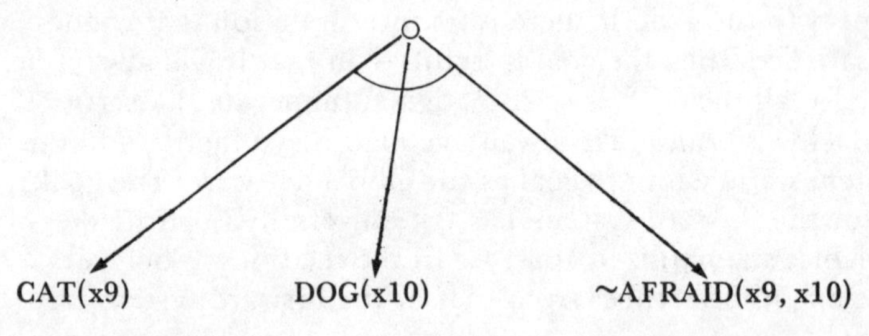

Figure 8.8 *An initial global database for a backward system*

The rules are applied, as before, but in the backward system it is the implied literal that is matched through unification (but in reverse). The result of applying the given rules so that the final database terminates in a consistent set of facts is shown is Figure 8.9. Consistency is checked by applying the unifying composition to all the unifers:

((Butch/x10), (x9/x3, x10/x2), (x10/x1), (x9/x6), (Tom/x9))

gives the two lists:

(Butch, x 9, x10, x10, x9, Tom)

and

(x10, x3, x2, x1, x6, x9)

The mgu for these two lists becomes:

(Butch/x10, Tom/x9, Tom/x3, Tom/x6, Butch/x2, Butch/x1)

thus showing that the result is consistent.

It is important in rule-based systems to choose the right form of the rule if a solution is to be determined. This choice assumes some knowledge of the problems to be posed and the method of solution. Rules can be manufactured from

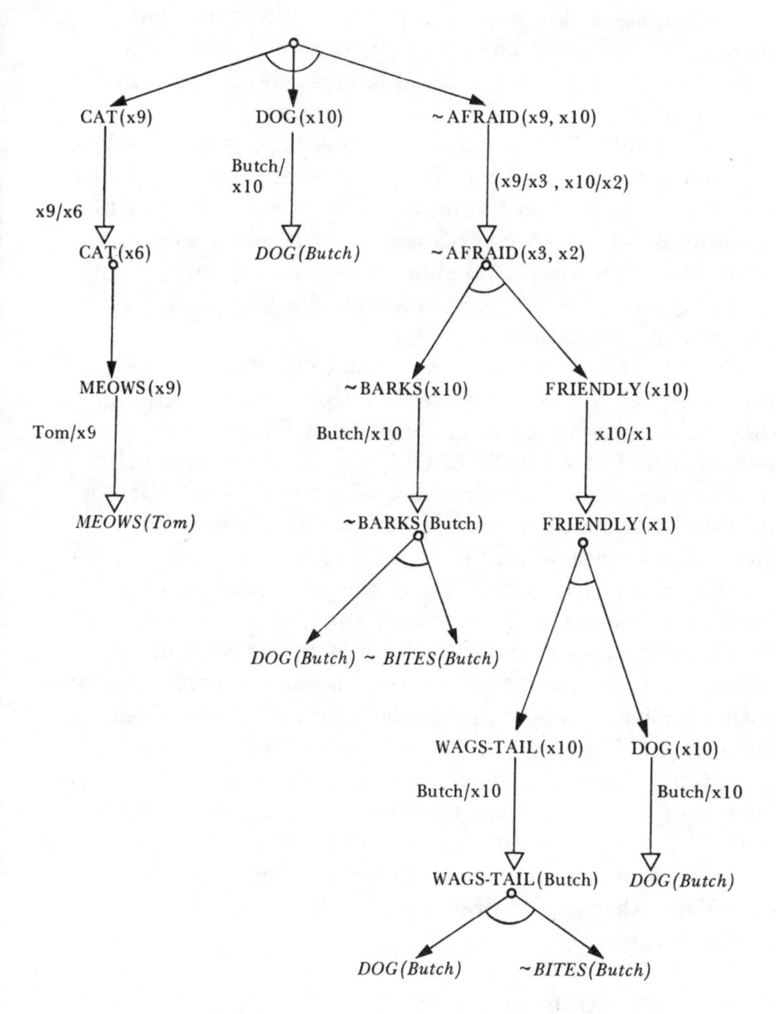

Figure 8.9 *A solution database for a backward system (adapted from Nilsson, 1980)*

other rules, and these will greatly improve the efficiency of the system at the cost of generality. Rule-based systems contain many hidden design problems.

The limits of machine reason

THE LIMITATIONS OF LOGIC FOR DESIGN

The formal approach to deductive reason has developed aside

from any considerations of computer architectures and business applications. The proof systems have mostly been implemented on the computer in languages that have their roots in mathematics rather than engineering (eg list processing or LISP). The working examples have at most only a few hundred facts and hypotheses, whereas business applications are concerned with millions of facts and only tens of hypotheses. The character of business systems has been formed through what is possible, desirable and controllable in a computer. The design approach to business systems reflects these constraints.

It is important in business systems to establish a coherent description of the task domain, and this requires a correct selection of predicates. Logic provides little guidance in making this choice [eg YELLOW(x) or COLOUR(x, yellow) or VALUE(colour, x, yellow) are all possibilities]. There is no distinction made between entities, structure and hypotheses. Each statement in logic is of equal importance and yet the roles of the statements in the process of reasoning can be very different. Recognizing this difference is part of the understanding that people have of their task domain.

Logic is indifferent to elicitation methods; there is nothing within logic about how knowledge of a task domain can be determined. Logic and the associated deductive reasoning provide no distinction between what is the case and what is possibly the case (this is a variation of the connection trap, see p. 94). The two statements 'dogs bark' and 'Butch is a dog' should not imply that 'Butch is barking'. The only conclusion that can be drawn is that 'Butch can bark but may not be barking'.

AN ERA OF A LOGIC PROBLEM

Extended Relational Analysis (ERA) has been developed to aid predicate selection through an elicitation method (see p. 68). This method ensures that the predicates conform by being both coherent and complete. There is also a natural distinction retained between the knowledge types. This distinction matches the client's view of the task domain, and this makes clear to him any explanations that are related to the form of the knowledge. The connection between ERA and large-scale business systems design (see Chapter 4) is well understood. The separation between 'actual' and 'possible' situations is expressible by both the choice of entity sets and the constraints defined between the entity sets. The advantages of ERA can be demonstrated through the example used to illustrate rule-based theorem proving (see Figure 8.4).

This example (described in Figure 8.4) suggests a task domain that can be expressed as a conceptual model. It was found helpful to use a variation of Brian Phillips' paradigmatic organization relationships to capture the structural hypotheses and facts (see p. 271). The major difference is that only True non-negated facts are recorded. Figure 8.10 illustrates the semantic functional dependency (Sfd) graph. Figure 8.11 gives both the intension and extension of the entity sets for this task domain. The entity sets EITHERTYPISA and ORTYPISA are to be interpreted as alternative sets of conjoined facts that are True, whereas TYPISA is the union of facts that are True in either EITHERTYPISA or ORTYPISA or both (eg 'Tom meows' is True in both cases).

The entities marked by (†) in Figure 8.11 are derived from the initial data given in Figure 8.4. In order to maintain consistency, the missing entities are determined by the following procedure:

ISA:= ISA ∪ (TYPISA * TYP) [Name, Class]
TYP:=TYP ∪ (TYPISA * ISA) [Act, Class]

The transformation rules R3.1 and R3.2 are used to derive facts, and they can be given in terms of relational operations, thus:

TRANSFORMATIONAL RULES (R3.1, R3.2)
BEGIN R3.1
TEMP1: = ((EITHERTYPISA | Act='Wags-tail')*(ISA | class='Dog')) [Name],
·*· ((VARACTS | Act='Friendly') [Act])
TEMP2: = ((ORTYPISA | Act='Wags-tail')*(ISA | Class='Dog')) [Name],
·*· ((VARACTS | Act='Friendly') [Act])
TEMP3:= TEMP1 ∩ TEMP2;
TYPISA: = TYPISA ∪ ((TEMP1 ∪ TEMP2) – TEMP3)
EITHERTYPISA:= EITHERTYPISA ∪ (TEMP1 – TEMP3)
ORTYPISA:= ORTYPISA ∪ (TEMP2 – TEMP3)
END R3.1
BEGIN R3.2
EITHERTYPISA:=
EITHERTYPISA
– ((EITHERTYPISA | Act='Friendly')
.Name * Name.
(EITHERTYPISA | Act='Barks' *or* Act='Friendly')) [Name]
·*· ((VARACTS | Act='Frightening') [Act])

```
ORTYPISA:= ORTYPISA
                – ((ORTYPISA|Act='Friendly')
                .Name * Name.
                (ORTYPISA|Act='Barks' or Act='Friendly'))[Name]
                ·*· ((VARACTS|Act='Frightening')[Act])
END R3.2
```

Thus, for rule R3.1 the first application will proceed as follows:

```
        TEMP1  [Name, Act]
                 —     —
        TEMP2  [Name, Act]
                Butch Friendly
        TEMP3  [Name, Act]
                 —     —
       TYPISA  [Name, Act:]
                Butch Barks
                Butch Wags-tail
                Tom Meows
                Butch Friendly
 EITHERTYPISA  [Name, Act:]
                Butch Barks
                Butch Bites
     ORTYPISA  [Name, Act:]
                Butch Wags-tail
                Butch Friendly
```

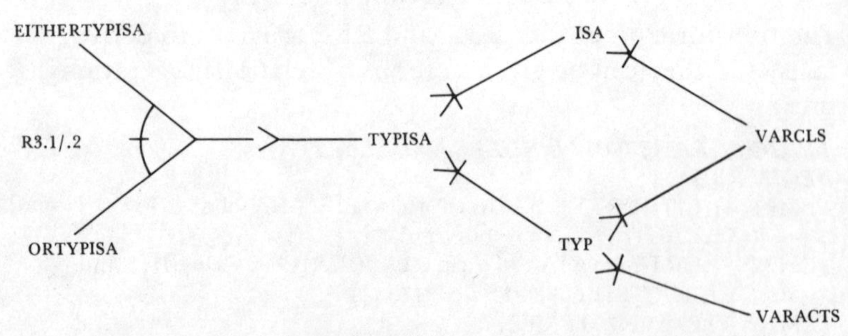

Figure 8.10 *The Sfd graph for the dog and cat problem*

The transformational rule R3.2 can then be applied to determine who is frightening to whom. The solution to this problem is very straightforward compared with the rule-based technique that was derived from the resolution principle. Rule-based systems perform the same tasks one inference at a time, whereas the relational operations deal with a complete group of inferences in a single step.

FACTS (R1.1, R1.2, R1.3)					
EITHERTYPISA [Name, Act:]		TYPISA [Name, Act:]		ISA [Name: Class]	
Butch	Barks	Butch	Barks	Butch	Dog
Butch	Bites	Butch	Bites	†Tom	Cat
		Butch	Wags-tail		
ORTYPISA [Name, Act:]		Tom	Meows		
Butch	Wags-Tail				

STRUCTURE (R4.1, R4.2, R4.3, R4.4)					
TYP [Act, Class:]		VARACTS [Act: Act2]		VARCLS [Class: Class]	
Meows	Cat	Barks	Noisy	Dog	Animal
†Wags-tail	Dog	Wags-tail	Act	Cat	Animal
†Barks	Dog	Meows	Act	Bulldog	Dog
†Bites		Friendly	Act		
		Frightening	Act		
		Noisy	Act		
		Bites	Act		

Figure 8.11 *The intension and extension of the dog and cat problem*

The concepts of VARACTS (variety of action) and VARCLS (variety of classes) explicitly define familiar hierarchical structures. TYP (ie typical) indicates that these actions (Act) are characteristic of these animals (class) but not always observable. ISA assigns the particular animal to its class whereas, TYPISA gives the current alternative situations (EITHERTYPISA, ORTYPISA). The transformation of the extension of the conceptual model is controlled by the rules and the procedures for maintaining Sfd consistency. Thus, when the facts change or new facts are added the extension can automatically derive further facts according to the given hypotheses.

THE BOUNDARY OF LOGIC

This chapter has presented a brief introduction to formal methods of reasoning. There are many areas of logic that have not been mentioned and there is a multitude of issues being researched. It is thus with caution that limitations are discussed.

There is no suggestion in formal reasoning that the methods simulate any physiological function. However, there is an underlying belief that the formal methods reflect some characteristics of rational argument. In particular, the methods are *deductive* in that a hypothesis (such as P→Q) and evidence (such as 'P is true') will generate the conclusion ('Q is true'). The alternatives to deduction are 'abduction' and 'induction' (not to be confused with mathematical induction) (see p. 139) both of which are such that conclusions do not necessarily follow from the evidence. Induction is invalid and abduction is characterized by theory formation which by its nature cannot be valid (eg the Lakatos Programme).

The methods described in this chapter and others also used for artificial intelligence are deductive because all the arguments are valid and valid arguments have been shown to occur only within a deductive framework.

Conclusions drawn from a deductive system are 'closed' in the sense that no *new* 'knowledge' of the world can result from a deductive system. The 'knowledge' may *seem* new in that an individual may be startled by the results of applying the methods, but this only reflects the limitations of human reason. The results cannot be *new* 'knowledge' since the links between the initial facts and these results are tautologies. What is new, of course, is the perspective, and the consequence of applying valid arguments is to observe the same knowledge from different points of view. Consequently, a machine that is uncoupled from the world will discover no new knowledge about the world, although it may be able to present new perspectives. Some of these new perspectives may eventually stimulate the clients to form new theories about the world.

CHAPTER 9

Applied Epistemics

> The Total Perspective Vortex derives its picture of the whole Universe on the principle of extrapolated matter analyses. To explain – since every piece of matter in the Universe is in some way affected by every other piece of matter in the Universe, it is in theory possible to extrapolate the whole of creation – every sun, every planet, their orbits, their composition and their economic and social history from, say, one small piece of fairy cake.
>
> *Douglas Adams (1980)*
> *The Restaurant at the End of the Universe*
> *Pan Books*

Introduction

The term applied epistemics has been adopted to refer to the techniques and methods for communicating understanding via stored knowledge. Recently it has been extended to include computer systems that embody models of competence and, in particular, expert systems (Addis, 1981, 1982d).

THE RANGE OF EXPERT SYSTEMS

One of the first practical expert systems was DENDRAL (Buchanan and Feigenbaum, 1978). Work on DENDRAL began in 1965 at Stanford, and its aim was to enumerate plausible structures (atom-bond graphs) for organic molecules from the information provided by analytic instruments and user-supplied constraints. The many possible structures to be explored require the use of search in order to consider exhaustively all the possible solutions. The domain searched is not an explicit set of known organic structures but an implicit set derived from a theory of how such structures are formed. This approach is formal and strictly non-human since it represents an intellectual ideal of how to achieve a solution which is beyond human capability. The excellent performance of DENDRAL was achieved by a symbiotic relationship between the competence represented within the system and the natural performance of the operator in guiding the system through its search task.

There have since been many other expert systems springing from DENDRAL. In the drive to create expert systems to model human thought processes, the medical consultant MYCIN (Shortliffe, 1976) and the geological expert PROSPECTOR (Duda *et al.*, 1976) were specifically designed to represent and explain all their reasonings in human terms in the sense of presenting the sequence of rules used to achieve a conclusion. In both these cases, the principle of generating vast numbers of solutions from some theoretical model is inappropriate. Also, it is assumed that no physician would (or should) consider using a proposed therapy and no geologist would recommend expensive mining operations unless each could justify the conclusions of the relevant expert system. The generation of an answer in a way that cannot be followed and assessed by the user was considered irresponsible. This pressure to simulate humanly recognizable reasoning produced several other systems. Many of these systems employed EMYCIN (Empty MYCIN) which captures the framework of a general expert system. It only requires the addition of knowledge.

Concurrent with these latest variations on the Stanford theme are systems developed directly from pattern recognition (Fu, 1958) and information retrieval (IR) techniques (Addis, 1982d; Oddy, 1977). For example, the Computer Retrieval Incidence Bank (CRIB) system (see Chapter 6) searches an accumulation of symptom patterns gleaned from many sources. It is the use of simple heuristics to provide feedback which differentiates this kind of diagnostic aid from a simple information retrieval system. The feedback is not just the presentation of the set of possible solutions but it also gives guidance on how best to refine the selection specification. At this level of interaction the appearance of the CRIB System is similar to that of rule-based systems. However, the nature of the knowledge retained and the capability of each system are subtly different. The CRIB System is biased more towards the accumulation and use of (observed) factual knowledge as opposed to developing a competence model of a human skill (cf Feigenbaum, 1980).

THE HUMAN FACTOR

The principal feature of expert systems is that they deal with problems whose specifications are complex, so that non-human methods can often be used to advantage. Complex problems usually lead to vast numbers of possible solutions, each of which must be examined with care – an assessment task that is often well beyond normal human capability. For example, DENDRAL is concerned with determining the molecular structure of organic compounds from the results of a mass spectrometer, where the problem is that the molecule gets broken down into untidy lumps (molecule fragments) which are recorded by the instrument. Consequently, the many possible structures that have to be explored require formal searching techniques based upon the theory of graphs in order to consider exhaustively all the possible solutions. Each solution is examined to see if it is a potential candidate for the particular spectral response of the molecule under test. Another example of non-human problem solving is MACSYMA (Bennet *et al.*, 1979). This is a program for evaluating integrals in mathematics. MACSYMA uses the Risch algorithm, which is never normally employed by mathematicians because of its complexity. However, the Risch algorithm is considered to be the ideal approach for evaluating integrals and consequently reflects a high standard of non-human competence. The competence described here is the formal method that generates solutions, and it is a model of the total set to be explored.

The medical consultant MYCIN (Shortliffe, 1976) and geological expert PROSPECTOR (Duda *et al.*, 1976, 1980), on the other hand, are specifically designed to represent and explain all their reasonings in human terms. The requirement was to provide a model of solution generation that is dependent upon a balanced mixture of facts and rules for generating potential solutions (Michie, 1979). This is achieved by gleaning understandable rules of diagnosis from the human experts and providing a control structure which can use these rules directly. This technique is also used in the most part by both DENDRAL and MACSYMA. The control structure will not only draw inferences from these rules but can calculate some measure of certainty concerning the final conclusion. The significant principle behind this measure of certainty is that it is not based upon traditional probability theory, which relies upon gathering statistics, but is intended to model an expert's 'belief' or 'disbelief' about a particular hypothesis given the accumulating evidence.

The control structure that handles this evidence, although more understandable to the user, does not in fact represent the manner in which human experts reason. The systems INTERNIST (an expert system for internal medicine; Pople, 1977), PSYCO (diagnosis of dyspepsia; Fox *et al.*, 1980) and PIP (concerned with oedemas and renal disease; Pauker *et al.*, 1976) have been developed specifically to mimic the human diagnostic procedure.

The main characteristics of MYCIN, PROSPECTOR, INTERNIST, PSYCO and PIP are goal-directed reasoning, sensitivity to patterns of symptoms and the abandonment of probabilistic (ie statistical) reasoning. MYCIN and PROSPECTOR will reason backwards from the goal (ie to determine the bacteria or ore deposit) to subgoals and then sub-subgoals in a way that is distinctly non-human. It is a form of reasoning through justification. PSYCO, on the other hand, attempts to reason in a forward direction in a way more akin to the observed behaviour of clinicians. The importance of goal-directed reasoning in either direction is that the user perceives a coherence in the questions being asked. The danger of the rule-based systems is that each answer to a question widens the circle of potential goal hypotheses that have to be investigated by the system, with the consequence that the system can ramble on long after a human expert would have come to a decision.

KNOWLEDGE ACQUISITION

All the systems mentioned above depend upon gathering knowledge in the form of rules (the 'production' rule technique). The advantages of these rules are that the knowledge concerning the particular area of expertise (competence) can be expanded incrementally and that it is in a form easily expressible by an expert. Associated with MYCIN, for example, is the program TEIRESIAS (Davis and

Buchanan, 1977), which collects new rules from the expert, checks the consistency of the rules and follows chains of reasoning under the control of the expert in order to discover inadequate or inappropriate rules. An equivalent program, Meta-DENDRAL, is linked with DENDRAL. TEIRESIAS is quite subtle and intricate in its operation, and this is achieved through a set of meta-rules used for collecting the diagnostic rules for MYCIN (see Chapter 1, Figure 1.3). These meta-rules are concerned with the general outline or form of the diagnostic rules and, in particular, which kind of diagnostic rules are associated together. They represent a generalization of the diagnostic rules which can be used to guide the MYCIN search strategy (ie only examine rules of a certain type under special circumstances). This principle of controlling the tracing of inferences at an abstract level before plunging into detail has been principally employed by ABSTRIPS: a general planner for a robot system at SRI (Sacerdoti, 1974).

The problem with rule collection is that it can be a long and tedious business. More easily collected are examples of symptoms that lead to a diagnosis; in the case of fault-finding in machines as opposed to the medical treatment of illnesses, these examples are often readily available as fault reports. Machine fault reports are frequently inadequate, incomplete, inconsistent and out of date, simply because the effort of maintaining them is labour intensive. The incentive for engineers or doctors to give accurate and complete reports is lacking because the usefulness of these records in their original state is minimal. However, provided a system will act as an automated clerk to analyse and correlate these reports to make them useful, then the quality of reporting should radically improve. A technique used in information retrieval for evaluating the usefulness of terms for retrieval has reappeared in the form of automatic 'induction' (Sparck-Jones, 1974). Work on automatic induction began at Stanford University with the objective of deducing a set of diagnostic rules from a database of examples that can be used by an expert system (Quinlan, 1979). The method of rule induction is similar to that used by the Rapid Action Fault Finding Library Enquiry System (RAFFLES) (see p. 252) to produce a fault-finding guide (Addis, 1978; 1982b).

The CRIB system, on the other hand, has been designed to

work directly on a database of examples without the need to deduce a set of rules (Addis and Hartley, 1979; Hartley, 1984). In this case, there is no attempt at generating a solution, except through unification. The CRIB system uses a few simple prefixed meta-rules under the control of the user to 'home-in' on a matching 'fault'. The user can draw on his own experience of the situation to guide the system to search for faults with particular attributes. In order to save on-line computer power the RAFFLES system precompiles the CRIB process by generating an optimum fault-finding guide (see p. 252). This achieves the same objectives (ie finding a matching fault) but only involves minimum on-line computer processing. The weakness of the initial RAFFLES implementation is that the user lies outside the diagnosis activity. The user is thus driven as a passive observer. This is a disadvantage of most expert systems since it does not fully exploit the user's local knowledge or performance ability. The main stumbling block is aways the problem of accepting descriptions of symptoms from the user in some specialized and restricted form of English.

Both MYCIN and PROSPECTOR have been deliberately structured so that their representations of rules are embedded in an organization suitable for 'natural' language dialogue. MYCIN converts the English form of the rules into a set of standard relations with a fixed set of attributes. This allows direct manipulation of the 'meaning' of the rules by the inference-making control structure, makes allowance for the context by constructing a 'context tree' from the conversation and uses this context tree to modify the generation of English to the user. However, the English to internal representation is weak. The internal representation to English mapping is used for new rule generation where checks can be made by regenerating the English from this internal representation.

PROSPECTOR employs semantic nets (see Chapter 10) to represent the rules. The intention of the designers and the potential of the system is to have free-form English conversations. This representation of rules in standard form is nevertheless important in that it is open to manipulation, transformation and merging, thereby providing a strength in the reasoning capability and subtlety in the interaction that is not present in any text-based system. Many systems rely upon segments of stored text that can be retrieved through keywords. However, to simply store text that is merely hooked into the system via keywords or other associated attributes is to create a system which is wide open to misinterpretation by users. There is always the possibility that the text, which is meant to 'stand in for' the keywords and the keywords' effect on the control structure, will imply more or suggest something different to that which was intended.

Uncertain knowledge

THE PROBLEM DOMAINS

Statistics and pattern recognition techniques as applied to medical records have proved to be unsatisfactory. However, diagnostics that involve objective tests have proved tractable to these techniques (de Dombal *et al.*, 1974). Initial trials to determine treatment within a confined area of diseases from a large database of examples have had some success (Nordyke *et al.*, 1971) but generally such an approach is of little use. The problem lies in the social aspect of medicine and medical treatment that reflects the continuous modification of a defined competence. Not only are medical records inaccurate but diseases themselves change from year to year. To compound this problem, the usefulness of disease categories is not permanent but is a function of changing treatments and styles (Edwards, 1972). Thus, any statistical approach that depends upon collecting material over a long period is doomed to fail because of the many combinations of patient types, diseases and treatments. This variation would demand vast quantities of examples to give statistically sound results.

MYCIN accepts that no single 'cure' for any bacterial infection can be recorded since combinations of bacteria may exist (and often do) within a patient who may be sensitive to certain drugs. Thus, MYCIN is designed first to find out which bacteria exist, then to determine the patient's sensitivities. From the physical attributes (weight and age) of the patient MYCIN can calculate the dosage of a selected minimum set of drugs (which do not affect the patient) to combat the bacteria. Although MYCIN is concerned with only a few bacteria (about 100), there is a large number of possible combinations of bacteria, patient sensitivity and drugs. There are further complications as a result of the progress of a disease and its different manifestations after past drug therapies.

The CRIB (and RAFFLES) technique can only respond with a tried and tested 'cure' but the problem domain is different. Although machine faults may interact with the software to produce different symptom combinations for the same fault, this interaction is both discrete and limited. It is discrete in that every machine is equivalent and symptoms are more often than not precise. It is limited in the sense that most conditions can be reproduced exactly so as to retrace events that led to failure. This can never be done in the diagnosis of diseases where every patient is unique, symptoms are vague and diseases change. Machine (or system software) faults will often accumulate into 1000 or 2000 different faults before new designs or modifications alter the object under test.

Further, old designs are often kept going long after the engineers (or software specialists) have moved on to different areas of interest. A more rigid or structured approach can thus be used in these cases, whereas the rule-based method may well require too much processing to be practical.

MODELS FOR INEXACT REASONING

A problem arises because most of the descriptions of the world by people depend upon linguistic performance and normative values. It is tempting to suggest that people are being inexact, vague, woolly or fuzzy but the language employed merely reflects the world as it is perceived. To say 'the girl was near me' is a more appropriate statement of fact than 'the nearest point of contact of the girl was 5.362 cm.' In the repair and maintenance of man-made machines most facts can be given unequivocally, but in the treatment and care of people or the divining of ores from geological observations the facts cannot be so precise.

The essence of the method that can cope with inexact facts within a computer program can be found in both Decision theory and Fuzzy Set theory. Fuzzy Set theory assigns a number (that ranges from 0 to 1) to the elements of a set called a membership value. The number indicates the degree of membership an element has to that set (0 = not a member and 1 = definitely a member). In MYCIN a solution for dealing with uncertain observations has been developed under the title 'a model of inexact reasoning'. This model uses equivalent equations and performs a comparable task to that of Fuzzy Set theory (Zadeh, 1965). The membership values in MYCIN are interpreted as Certainty Factors. However, there are further constraints on these membership values in MYCIN. For example, an object could be described as oval, round, square or rectangular in appearance, with no certainty that it is any particular shape. However, it may be more oval than round and more rectangular than oval. This indistinction of category is quantified by assigning a number between 0 and 1 to each description. Thus, the object's shape can be described as [(rectangular, 0.6) (oval, 0.3) (round, 0.1) (square, 0.0)] where the numbers indicate some measure of certainty for each classification. Since the object has to have some shape, the sum of the assigned numbers is expected to be not greater than one. However, if these numbers are membership values, then this restriction is not necessary. Thus, unlike a normal database, an attribute can be multi-valued and indecisive.

Objects are not normally identified according to a single attribute, and even if each attribute had a definite value the classification of such an object can also be indistinct. Linear equations are provided with arbitrary thresholds which allow the uncertainty of attributes describing an indistinct object to arrive at a useful decision. Objects for MYCIN are bacteria whose attribute values are derived from both clinical and laboratory data. The nature of the problem domain is such that supportive evidence for one bacterium does not preclude the existence of others.

Natural language provides tools for making subtle distinctions along attribute dimensions. A colour can be simply 'red' but it may also be described as 'a sort of' orange-red. This state of affairs may be represented within a computer program as [Colour(Red 0.7) (Orange 0.3)] or (Colour 6000) where 6000 is the wavelength. The term 'sort of' can be predefined for each colour but the range of possibilities can be extremely large. Shaket (1976) gives a technique that employs a membership function that converts physical values to certainty (membership) values. These values can then be modified by transformations associated with linguistic hedges —such as 'very', 'too', 'rather', 'most', 'less', 'sort of' etc – which effectively cause a shift in fuzzy set values to accord with human expectations (see p. 234). However, it has been suggested that using an assessment of hedges by adding up the pros and cons could also be as effective as a numeric model of English descriptions (Fox, 1981).

Two models of inexact reasoning will be described. Bayes' rule is the basic decision mechanism employed in a number of environments, and it is thus worth understanding in detail. The Fuzzy Semantic model is interesting because of the simple manner in which complex concepts can be expressed and acted upon.

The Bayesian Theory of Classification

The Bayesian Theory of Classification is devised from the elements of Decision theory. Central to the theoretical treatment of Decision theory is the loss function $\lambda(i/j)$, where i is any one of R hypotheses and j is any one of the same set of hypotheses. This function represents the loss (in terms of units) incurred when the decision is made that the state of the world is i when, in fact, it is j. In terms of pattern recognition, this represents the loss that occurs when the machine places a pattern that actually belongs to category j into category i.

One of the simplest expressions of this function is:

$$\lambda(i/j) = 1 - \delta ij \qquad (9.1)$$

where $\delta ij = 1$ when $i = j$ (Kronecker delta function).

$= 0$ otherwise.

This states that the loss when wrong will be 1 unit and that there will be no loss when correct. If a machine classifies patterns such that the average value of $\lambda(i/j)$ is minimized, the machine is said to be optimum.

A pattern of evidence can be represented as a vector X in the feature space, where the features are observations that represent the different items of evidence. The probability of a pattern of evidence X implies a hypothesis i, where i is one of R hypotheses, will be represented by the symbol $P(X/i)$, and the probability of the ith hypothesis will be given as $P(i)$.

Now, it is likely that the features chosen are not perfect

for distinguishing a state of affairs given a pattern of evidence, and there will be an overlapping of the hypotheses' boundaries. The probability of a hypothesis j given a particular evidence pattern X will be represented by P(j/X).

If the decision i is made (the basis for making this decision is immaterial at this point), then the conditional average loss L(i/X) will be:

$$L(i/X) = \sum_{j=1}^{R} \lambda(i/j)\, P(j/X) \tag{9.2}$$

This will be the average loss (in units) for making this decision given the particular pattern X. Combining Equations (9.1) and (9.2):

$$L(i/X) = \sum_{j=1}^{R} (1-\delta ij)\, P(j/X)$$

and this becomes:

$$L(i/X) = 1 - P(i/X) \tag{9.3}$$

Therefore, to minimize the expected loss for a particular decision i, *i should be chosen so as to maximize P(i/X).*

Now P(i/X) can be estimated by using Bayes' rule, which is:

$$P(i/X) = \frac{P(X/i).P(i)}{P(X)} \tag{9.4}$$

Now if X is described by the features x1, x2, x3, . . . xd . . . xD, then:

$$P(X/i) = P(x1/i).P(x2/x1, i).P(x3/x1, x2, i) \ldots \text{etc}$$

and:

$$P(X) = P(x1).P(x2/x1).P(x3/x1, x2) \ldots \text{etc}$$

therefore:

$$P(i/X) = P(i).\frac{P(x1/i)}{P(x1)}.\frac{P(x2/x1, i)}{P(x2/x1)} \ldots \tag{9.5}$$

This would take considerable computing power if X has many dimensions, and there would need to be a large number of examples of each i to ensure the correct determination of some of the more obscure probabilities.

If the transformations on the measurements (observations) are chosen so that the features are independent of each other, then Equation (9.5) can be greatly simplified:

$$P(i/X) = P(i).\frac{P(x1/i)}{P(x1)}.\frac{P(x2/i)}{P(x2)}.\frac{P(x3/i)}{(Px3)} \cdots \text{etc}$$

This means that the transformation must do the work of correlation.

If logs are taken and log (P′(i/X)) is considered as a *discriminant function* G(i/X), then:

$$G(i/X) = \sum_{k=1}^{D} \log (P(xk/i)) - \sum_{k=1}^{D} \log (P(xk)) + \log (P(i))$$

Since $\sum_{k=1}^{D}$ log P(xk) is a constant for each of the k hypotheses:

$$G'(i/x) = \sum_{k=1}^{D} \log P(xk/i) - \log (P(i))$$

This is an extension of Bayes' rule in log form. It is a simple linear equation where the impact of all the evidence is the log sum of the impacts made by each observation separately. The only requirement is that the observations must behave as though they were independent. The individual elements (log P(xk/i)) can be calculated for any 'pattern of evidence', and if there are several hypotheses to choose from then maximum G′(i/X) is selected.

Fuzzy semantics

The natural language understanding system SHRDLU devised by Winograd (see Chapter 10) used a toy world of blocks where all the objects were equally distinct. Thus, a red-orange block was no closer to being a red block than to being a black block, and a long plate was as separate from a short strip as from a cube. The dissertation 'Fuzzy semantics for a natural-like language defined over a world of blocks' by Shaket (1976) describes a fuzzy set semantics for a limited English-like language (in which red-orange is near to red and so on). The system accepts instructions in the form of simple indicative sentences and demonstrates its 'understanding' by indicating the block which is the intended referent. The work is based on the 'principle of maximum meaningfulness' (Goguen, 1976), and is robust in the sense that small changes in the instructions or in the state of the world do not disturb it.

The objects in the fuzzy blocks world are specified by height, width, length, colour and position dimensions. Thus, an object B3 would be internally represented as:

B3:	*W*IDTH	22 (cm)
	*D*EPTH	17 (cm)
	*H*EIGHT	15 (cm)
	*C*OLOUR	550 (wavelength)
	etc	

Note that no attempt has been made to classify the object at this stage since its class membership will depend upon its relationship to other objects in the context of the request.

The definition of objects relies upon two membership functions, which give the degree of membership (truth value) to a class according to the value of a particular measurement (x). These membership functions are:

$$f(x) = 0.5 + \frac{1}{\pi} \tan^{-1}\left[\frac{x - k1}{k2}\right]$$

and:

$$g(x) = \exp - \left[\left(\frac{x - k1}{k2}\right)^2\right]$$

which are illustrated in Figure 9.1 (a and b).

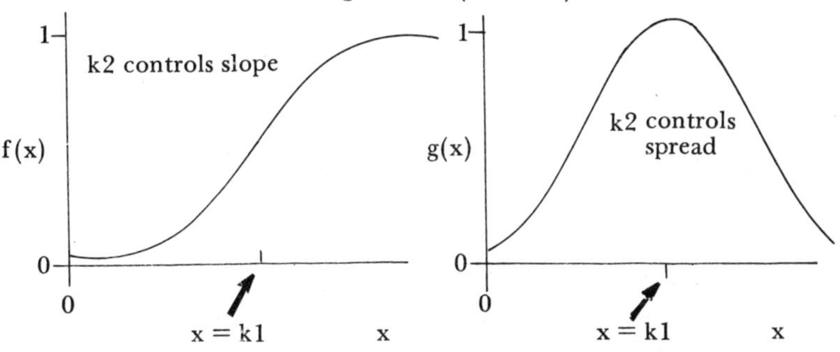

Figure 9.1 *The two functions that map measurement to membership: (a) f(x) reaches 0 and 1 at −∞ and +∞; (b) g(x) has a centre value equal to 1 and reaches 0 at ±∞*

Associated with each noun in the dictionary, there is a membership function constructed from f(x) and/or g(x), where x is some simple function of the dimensions of objects. Thus, a cube is defined as:

$$\text{CUBE (W, D, H)} = \text{f1}\left[\frac{\text{MID}}{\text{MIN}}\right] \quad \text{f1}\left[\frac{\text{MAX}}{\text{MID}}\right]$$

where:

MAX = maximum side length (W, D or H)
MIN = minimum side length
MID = remaining side length
k1 = 1
k2 = 0.3

A plate, on the other hand, is defined as:

$$\text{PLATE (W, D, H)} = \text{g1}\left[\frac{\text{MIN}}{\text{MID}}\right] .\text{g2}\left[\frac{\text{MID}}{\text{MAX}}\right]$$

	g1	g2
k1	0.04	0.8
k2	0.15	0.5

Adjectives are modifiers which shift and reshape the characteristic membership function so that it uniquely refers to the object intended. Membership of an object in the noun group (group of words describing an object) is found by multiplying and normalizing the values given by the noun membership function and the adjective membership function. For example 'elongated' is defined as:

$$\text{LONG (W, D, H)} = \text{f2}\left[\frac{\text{MAX}}{\text{MID}}\right] \quad \text{such that k1} = 4.25$$

and the 'elongated plate' membership function becomes:

$$= \text{LONG (W, D, H)} \ . \ \text{PLATE (W, D, H)}$$

Adjectives and adverbs are often preceded by words such as: very, too, rather, most, less, sort of etc. These words belong to a structure group called hedges. A shifting effect, similar to the one performed by adjectives, can be brought about by a transformation on the truth values (result of an applied membership function) of objects. The hedges can then be interpreted as transformations which shift attention to a different point from that represented by the largest truth value.

Determiners and numbers are structure words which specify how many objects are indicated and how definite the choice should be. The determiners, apart from number, also specify the expected relationships between the membership value of the object sought and the membership values of other relevant candidates. The difference between 'a' and 'the' is that 'a' requires any appropriate member of the fuzzy set, whereas 'the' requires a distinct member.

There are two kinds of threshold values which select objects. The first is the 'absolute threshold' used for the determiner 'a':

$$\text{THA} = (0.5)^{k}$$

where k is equal to the total number of multiplications performed in obtaining the result. The second is the 'ratio threshold' used for the determiner 'the' which sorts the list of membership values into decreasing order, divides each value by the following value and then tests the Nth ratio. If the Nth ratio passes the threshold 4/3, where N is equal to the number part of the noun group (and N = 1 if no number given), then this object is selected.

The determiners 'a' and 'the' operate differently from each other when dealing with relationships. Consider the two sentences:

1. Find a cube which is near *a* red plate.
2. Find a cube which is near *the* red plate.

In the first case (a), there is a non-definite noun group so that all combinations of CUBE and RED PLATE have to be evaluated by the relation NEAR. This will generate a matrix of truth values. In the second case (b), the RED PLATE can be considered as a pivot to which NEARness of all cubes is tested.

The system can only deal with imperative sentences of which the subject of the sentence is an implicit 'you'. The only verb is FIND, and the object is a noun group (NG). The role of this clause is an indicative code which is being modelled by the interaction of membership functions and fuzzy set operations. The noun group has a very strict syntactic structure in English, which can be represented as:

DET NUM (HDG* ADJ)* *NOUN* QUALIF*

where the components which have to be present are set in italics and those components that may be repeated are denoted by an asterisk (*), so 'FIND THE VERY BIG RED PLATE' has the structure:

(S(VG FIND) (NG ((DET THE) ((HDG VERY) (ADJ BIG))
(ADJ RED) (NOUN PLATE))))

The logical connectives for fuzzy sets can(but not always) define the membership value of a set of conjoined phrases as the minimum value and the membership value of a disjoint set of phrases as the maximum value. Thus, the values of a 'big and red cube' could have been calculated as:

MINIMUM (big cube, red cube)

and 'a big or red cube' could be:

MAXIMUM (big cube, red cube)

In Shaket's program, each of the words BIG, RED and PLATE is a content word whose meaning is a fuzzy set over the space of object features. VERY is a structure word which modifies the fuzzy vector passed to it by BIG. These results are then multiplied together to form a new fuzzy vector over object space. THE indicates how many objects are to be chosen and in what way, and FIND makes the final selection. The process generally operates from right to left requiring very little parsing.

PROBABILITY AND MEASUREMENT

The allocation of membership values or the results of the membership functions to different objects according to their various dimensions is completely arbitrary. Even the fuzzy set functions used for hedges (which shift attention from the maximum truth values) are invented rather than discovered. A membership value can be considered (but not generally so) as:

> the probability that a particular object would be classified as a 'block', 'red' etc when presented to a large number of people. The functions f(x) and g(x) would then represent the change in this probability for a change in a particular dimension.

Membership functions as used by Shaket could then be

determined in an almost identical way to that of discrimination functions used for pattern recognition (ie Bayes' rule). Further, it is possible to discover *by experiment* the fuzzy set function employed for hedges and the manner in which adjectival functions interact with membership values of primary objects.

Entities must have a fixed set of measurements (eg width, height etc,) if they are to be compared. In practice, this check is not necessary since the absence of applicable measurements results in meaningless statements as in:

'The tall moon' (no height feature attached to the noun)

THE RATIONAL ASSESSMENT OF ASSERTIONS

The application of Fuzzy Set theory and Decision theory captures some components of uncertainty. The assignment of probability or set membership to observations can be interpreted in ways not intended by the original definition. In order to explore a different view of uncertainty (or membership value) assignment to assertions, a simple view of the formal logic will be assumed. With this provision in mind, the rational assessment of uncertainty can be considered as a problem of *judgment* rather than a measure of 'truth' or 'probability' in some absolute sense.

It can be said that formal logic preserves truth and provides an exact means of deducing propositions from assertions. The inference mechanisms may be judged with respect to different models of truth assignment (eg to test validity and satisfiability), but the actual assignment of truth to the initial axioms is not considered a problem. It is accepted that the 'truth' of statements for the application of reason (eg as required by expert systems) is not clear due to the open-ended nature of most human tasks. To this end, various formal logics have been devised that manipulate uncertainty (and thus reflect the open-ended nature of the problem) in a manner analogous to that in which predicate logic preserves truth (Mamdani and Efstathiou, 1984). These *logics* of 'uncertainty' are not the concern of this section. They represent various options, that are well-defined, from which models of human reason might be derived. The question explored here is 'what is involved in the assignment of uncertainty to assertions before such logics can be used?' In order to examine such a question, it is necessary to draw upon the ideas presented by Stephen Toulmin (1958) in his book *The Uses of Argument.*

The rational assessment of an assertion requires formalities, but these formalities are not sufficient on their own to achieve judgement. The formalities usually involve a clearly recognized sequence of activities that starts with a claim, is followed by the evidence and finishes with a verdict. In logics that use concepts like Certainty Factors, the verdict of each primary statement from which conclusions are to be drawn is presented in the form of a number. The numbers and the associated statements are the base from which all further judgements are made. The numbers are manipulated to produce consequential verdicts on statements derived through the use of the logic.

How the base numbers are assigned, upon which these consequential verdicts depend, is never recorded, and the principles involved are inaccessible to any questions of validity.

When an assertion is considered to be possible, then this is a claim on our attention, and different claims are more or less serious. The degree of importance of the claim is dependent upon the context and the potential case for support. Some claims are an exercise in logic and are supported through the formal application of an accepted method. Other claims, such as base statements, have to appeal to alternative modes of judgment. Thus, a particular statement is possible in mathematics when there is no demonstrable contradiction. An axiom in mathematics cannot be questioned through mathematics alone. In a wider frame, and in other fields, possibilities are determined through different *criteria.* Criteria are the grounds required to justify an assertion, and these grounds are dependent upon the *field of argument.* Arguments belong to the same field if they are of the same *logical type*, where a logical type is a set of objects that shares a coherent domain of operations. Examples of different logical types are:

1. The proofs in Euclid's Elements.
2. The calculations for weather forecasting.
3. 'This animal lays eggs, has feathers, and is, therefore, a bird'.
4. 'The defendant was driving at 45 mph in a built-up area, so he has committed an offence against the Road Traffic Act'.

Each field will have its own criteria and *standards* from

which judgments can be made. However, even within a field there is a choice of criteria depending upon the required consequences. Economic decisions may be made according to certain measures such as 'maximum average wealth', but such criteria when pushed to the extreme may have undesirable results. Thus, criteria and standards are changed in order to achieve specific ideals.

Force is a field-invariant measure of the argument that supports an assertion. Force is the assessment of the validity of the assertion after the appropriate criteria and standards have been applied. Thus, a given statement is 'true' because it satisfies the criteria and standards required to achieve a purpose, and it will only be true given certain 'understood' restrictions. If these restrictions are breached, then the statement can become 'false' or at least suspect. The validity of such an assertion is dependent upon the limits of the imagination to stretch the boundaries of the context in which the statement is given. However, given a well-defined context, the argument for support has a force that can be judged in ordinal terms. Force can then be used to compare the two assertions of different fields of argument. *It is the force of the argument that supports an assertion*, and it is this force that is often summarized as an associated number (certainty factor or membership value). This number may then be manipulated by 'models of reason' such as fuzzy logic provided that the contexts of the assertions are compatible. Assertions made in different contexts, their forces recorded and employed to draw conclusions can produce obvious nonsense.

It could have been argued that there is a distinction between an assertion made as an axiom (base assertion) and an assertion derived through the application of formal reasoning. Thus, to treat them as similar is a mistake. However, it is not clear what the distinction might be. The only distinction that can be made is that the former is accepted without the extension of *the case for support*, whereas the latter has the case completely exposed. The potential case for support of base assertions must be available within some field of argument (which may be different from the formal mode it is 'fed to') in order that a number may be assigned. The case may be weak or strong but in any event these arguments should be capable of being pursued if necessary. There are some assertions that are true (or false) in 'all possible worlds' but these assertions do not constitute knowledge 'about' the world. They certainly cannot be doubted and they are not in question.

In summary, the force of the initial axioms can be expressed as numbers. The numbers are used so that assertions may be compared on a single scale. The numbers may be derived from arguments of different logical types to those which will eventually manipulate the axioms and their associated numbers. The criteria and standards from which the force is assessed remain isolated from the logic in which they are used. It is important that the assertions upon which a logic is to be employed are coherent with respect to the criteria and standards of the context. The criteria and standards depend upon the purpose behind both the distinctions being made and the choice of logic employed. The significance of these distinctions may be reassessed in the light of the conclusions arrived at by the action of the logic. It is the study of how these distinctions and choices of logic arise that will eventually lead to a description of what is involved in the numeric evaluation of base assertions.

MYCIN

MYCIN has been used as the principle model from which expert systems have been derived (Cendrowska and Bramer, 1982). An analysis of MYCIN in the terms developed in Chapter 3 illustrates the continuity of expert systems with other computer systems. The task domain constraints, as represented by the Semantic functional dependency (Sfd) graph, are used both for maintaining the relative truth conditions of the stored data and for generating search tasks from user-specified requests. A system that uses inexact reasoning is also concerned with gathering facts in order to produce a statement of the task domain as perceived by the user. To gather the facts correctly still requires update protocols synthesized from a model of the relationships between the facts. MYCIN retains fixed facts about bacteria and the symptoms normally associated with them. MYCIN also has information on the drugs and what sort of bacteria they can affect as well as the potential side-effects of the drugs on the typical patient. MYCIN constructs a model of the patient from the facts, and this model is in the form of a database confined to particular update constraints. Once the model of the patient has been achieved, MYCIN can then call upon special assessment procedures which use the characteristics of the patient defined within the database to recommend a therapy of a minimum effective set of drugs.

The primary task of MYCIN is to determine what significant organisms exist within the patient. To achieve this task there are two kinds of production rules of the form:

IF ⟨premise⟩ THEN ⟨action/conclusion⟩

The first kind of rule is concerned simply with the problem of collecting data from the consultant using the system. These data may be results of laboratory tests or observations made by the consultant. The second kind of rule is used to maintain data consistency. The update rules are normally IF . . . THEN rules but they can also be represented by an Sfd graph. The advantages of the Sfd graph are that it provides a uniform method of description that can be displayed in a single figure and it can lead to a reconstruction of MYCIN that can take full advantage of conceptual model implementation techniques.

Rules of both kinds can have premises which are indistinct (fuzzy), and associated with the interpretation of these rules is a procedure that can calculate a value representing the certainty of the conclusion (action). Premises consist of a list of conjoined conditions, and these conditions are concerned with statements of facts such as 'name of patient', 'identity of organism', 'drug allergies', 'locus of infection', 'result of a test' etc. Each of these facts can have a range of certainty that must be satisfied in order to be considered true. Examples are given in Table 9.1 and 9.2.

	fuzzy function	context	parameter	certainty range
1	It is definite	that the patient	is Jones	+1.0
2	It is known	that the organism	is streptococcus	+0.2 to +1.0
3	It is not definite	that the site	is blood	-1.0 to +1.0

Table 9.1 *Examples of fuzzy functions that do not form conditionals but describe attribute status*

	context	attribute	fuzzy relation	value	certainty range
4	The organism-1	morph is	same as	rod	+0.2 to +1.0
5	The organism-3	Gram	definitely is	Gramneg	+1.0
6	The organism-4	air	might-be	aerobic	-0.2 to +1.0

Table 9.2 *Examples of fuzzy functions that control conditional statements on clinical parameters*

Rules can be constructed using the kind of fuzzy functional statements given in Tables 9.1 and 9.2. Thus:

IF the organism-1 Gram is same as Gramneg
and morph is same as rod
and air is same as aerobic

THEN Conclude that organism-1 classification is Enterbacteriaceae
with a certainty of +0.8

If the database on the patient contained information shown in Table 9.3, then the result would be:

MINIMUM (1.0, 0.8, 0.6)
= 0.6

and this is true because it is in the range +0.2 to +1.0, which tallies with the fuzzy requirement 'same'.

context	parameter	value
Organism-1	Gram	((Gramneg 1.0))
Organism-1	Morph	((Rod 0.8) (Coccus 0.2))
Organism-1	Air	((Aerobic 0.6) (Facul 0.4))

Table 9.3 *An example of a fuzzy database with multiple values*

The result of 0.6 now interacts by multiplication with the certainty factor of the conclusion giving a value of 0.48. There are three English translations of the positive value (the negative value simply prefixes the translation term with not) depending on the range of this certainty value:

Strongly suggestive	(+0.8 to +1.0)
Suggestive	(+0.4 to +0.8)
Weakly suggestive	(+0.0 to +0.4)

The above example could, therefore, cause the program to respond with the tentative goal hypothesis:

There is suggestive evidence that organism-1 is Enterbacteriaceae (0.48)

Further evidence found by other rules for the same goal hypothesis can also modify this result by the formula:

$$C1 + C2 * (1 - C1) = \text{new certainty factor for hypothesis}$$

where C1 and C2 are the Certainty Factors of the final conclusions of the two interacting rules about the same hypothesis.

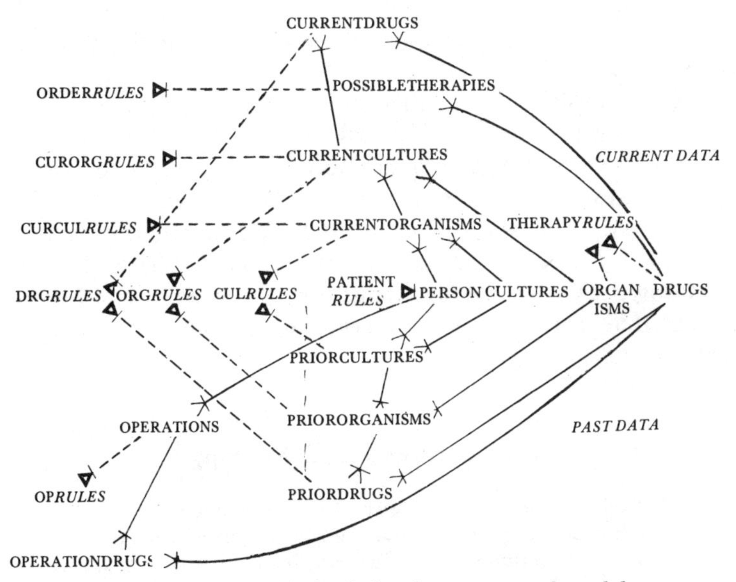

Figure 9.2 *The Sfd graph for the conceptual model component of MYCIN*

The Sfd graph shown in Figure 9.2 was determined from *Computer-Based Medical Consultations: MYCIN* by Shortliffe (1976). The descriptive information was incomplete, and hence various dependencies had to be assumed. The additions to the conceptual model are the rules associated with some of the relations. These local rule sets are shown at the end of a dashed 1-implies which is an abbreviation for the JOIN (see Figure 9.3).

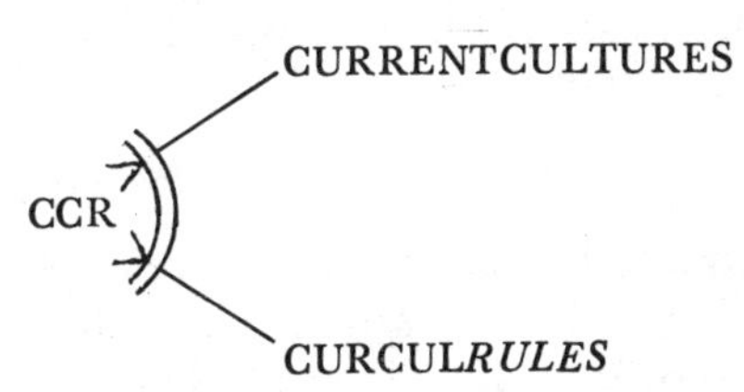

Figure 9.3 *Part of the Sfd graph showing the description of rule dependency*

When two relations such as CURRENTCULTURES and CURCUL*RULES* are JOINED together, then the relation CCR is the set of active rules associated with the specific cultures recorded in CURRENTCULTURES. Thus, every culture must trigger its appropriate set of rules when inserted and also make available those rules needed for the decision process.

The Sfd graph shown in Figure 9.2 together with the associated set of rules describes a system that controls the creation of a structured database about a patient. In addition to these rules for maintaining structure (entity set constraints), there is a further set of rules concerned with collecting information about the patient. The control of collecting the data flows back and forth between the two sets of rules. The primary data collecting rule is rule 092 which states:

IF:

(1) There is an organism which requires therapy.
and
(2) Consideration has been given to the possible existence of additional organisms requiring therapy, even though they have not actually been recovered from any current cultures.

THEN:

Do the following:
(1) Compile the list of possible therapies which, based upon sensitivity data, may be effective against the organisms requiring treatment.
and
(2) Determine the best therapy recommendations from the compiled list.

OTHERWISE:

Indicate that the patient does not require therapy.

In order to satisfy the premise of rule 092, other rules that have the conclusion:

'There is an organism which requires therapy'

are extracted and applied. In this case there is only one rule (rule 090) which satisfies this requirement, and this in turn leads to other rules which finally collect data about the patient. When this occurs, the rules concerned with maintaining the constraints are triggered since information about organisms cannot exist until there are cultures (see Figure 9.2). Further cultures cannot occur unless there is a person, and the first thing required to be known about a person is name, age etc. MYCIN will, therefore, print the relation name being created followed by an entity number which becomes an internal identifier of this entity (key). This is then followed by a request for attribute values. The user's responses are italicized.

PATIENT 1

1. Name: *J. Sample*
2. Sex: *Male*
3. Age: *60*

Other constraint rules are activated as the database under construction checks consistency and demands information from the user or searches its own database to maintain it. Once all the database requirements are satisfied, then the data collecting rules are allowed to proceed. Thus, MYCIN introduces four important additions to an information system:

1. Weighted multiple values of attributes.
2. Additional update rules associated with relations.
3. Meta-rules concerned with database creation.
4. An interrogation and search procedure that involves the construction of a model of the problem (patient) under consideration.

When MYCIN has satisfied all the consequences of the premise of rule 092, then data assessment procedures are activated (as in the CRIB system) to compile a list of possible drug combinations that will attack the organisms and are suitable for the patient.

Types of epistemic systems

Expert systems were derived mostly from work that was concerned with emulating human intellectual behaviour. The reason for this concern was to understand and study the mechanics of human thought processes. From this study it was believed that machines could be created that would have all the advantages of intelligence and insight without the human disadvantages of forgetfulness and error. The expert system was, therefore, meant to represent the complete and ideal expert within a narrow field of knowledge who would not forget, draw illogical conclusions, be biased or need sleep.

There are five different types of epistemic systems. The differences between these types are expressed as enhancements to computer data retrieval systems (see Table 9.4).

system type	enhancements and description
0	data files with on-line access at the *logical* interface
1	files with on-line access and the information may be *fuzzy* ie weighted multiple values
2	world constraints maintained automatically by update procedures, *conceptual* interface
3	fact *assessment* procedures and mechanisms to aid fact retrieval, rules for problem solving in the task domain
4	rules for collecting new facts (ie *meta-rules*)
5	mechanisms for *generating new rules* that can modify the task domain constraints (new model), adjust assessment procedures and provide new rules for collecting facts about different hypotheses

Table 9.4 *Enhancement table for epistemic systems*

An expert system can be viewed as an information retrieval aid for experts that involves update procedures and methods of communicating group practices. The major sources of confusion that can occur if expert systems are considered equivalent to a human expert are the distinctions between:

1. naturally occurring performance;
2. the formal representation of an abstraction of that performance;
3. the simulation of such a representation on a computer;
4. agreed ideal behaviour or practices.

Most artificial intelligence systems simulate the agreed ideal behaviour but there is a trend towards simulating the abstraction of naturally occurring performance as a model of human cognition. Within the realm of expert systems this path is taken in order to ensure that human decision makers can follow arguments that support any proposals made by the system (Michie, 1979). In this way, the responsibility of any action based on these proposals can remain squarely upon the shoulders of the experts and their novices. However, this trend weakens the usefulness of expert systems in that it restricts the strategies of problem solving to those within the users' intellectual grasp.

An alternative view can be taken which considers the expert system as a source of mutually agreed competence that is capable of providing non-human (opaque) solutions to problems. However, although the derivations of these solutions may often be beyond the capability of any individual to follow, the user will always be in the position to decide the relevance of such mutually agreed competence and should thus be placed in a position of controlling the search through an implicit maze of potential solutions. Such competence exists because there is a formal system of validation. However, such validation procedures may also be used to capture competence from performance. One such process is called knowledge refining.

Knowledge refining

SCARROTT'S CONJECTURE

It has been suggested by those holding an extreme view of artificial intelligence that human activity can be represented by some simple construction of rules. The opponents of this view have argued that human behaviour is so diverse that even within a restricted environment there is no simple framework which could adequately describe this behaviour. One activity that reflects and records something of the structure of human thought is a computer program. In the search for a measure of usefulness of computer storage, Scarrott (1973) (see also Bennett, 1975) suggested that communication theory (Shannon and Weaver, 1949) was inadequate for describing the human use of computer stores. It is true that for every additional bit added to a computer store, the number of possible states and hence, in principle, the representation capability is doubled. In the case of representing characters and numbers within a store, this fact is used within the limit of six to eight bits. However, for stores much larger than this it is necessary for people to structure their problem in a way that takes no cognizance of the potential channel capacity of the store. There seem to be other factors at work which are dependent upon the limitations* of the human user, and it is these limitations rather than entropy which govern the usefulness of the computer store.

Of interest here is not a measure of store utility but the restrictions that govern human behaviour. A clue to these restrictions can be found in the observations made by Zipf

*It could be argued that because there is *no* fixed class of structures that bind the human user, such as the character set, then the factors at work are also dependent upon the *un*limited possibilities of the users' constructions.

(1949). He showed that taking a histogram of word frequencies ranked in descending order (ie the most frequent word first etc) and plotting this on a log rank versus log frequency graph, the result is a straight line (see Figure 9.4).

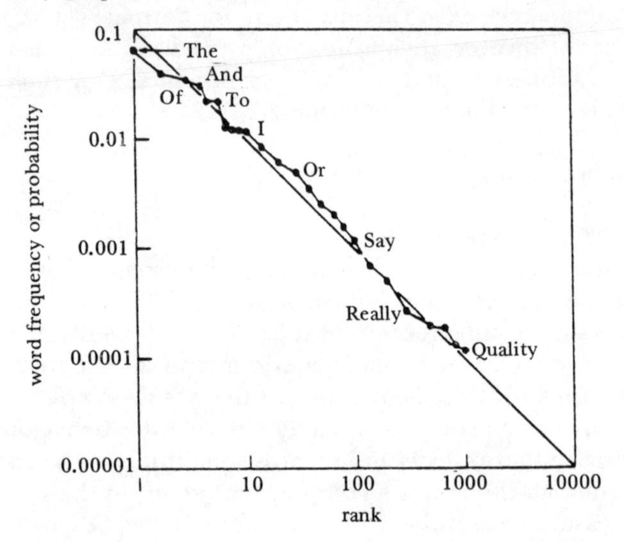

Figure 9.4 *Rank/frequency graph for English words*

Similar observations have been made for the frequency of address pointers in list processing (LISP) (Clark and Green, 1977), sizes of income (Pareto, 1897) and many other records influenced by the human need to structure activities. In particular, the distribution of the classification of diseases in medical diagnosis (Greenwood, 1972) and the use of keywords in document retrieval all obey Zipf's law.

The normal reaction to Zipf's law is that it requires no explanation since the effect is an inevitable consequence of the ordering of a random distribution. However, the ordering of a random distribution obtained from arbitrarily dividing a fixed set of objects into subsets produces the radically different Whitworth curve. Although theoretically derived (Whitworth, 1901), it can be illustrated by the rank/ frequency distribution of the use of individual letters in English (see Figure 9.5).

There have been many attempts to explain Zipf's law (Zipf, 1949; Fairthorne, 1969; Mandlebrot, 1977), and in the light of the Whitworth distribution some explanation is required. The clearest hypothesis is that Zipf's law is a consequent of the self-similarity of recursively defined structures (Mandlebrot, 1977). Such structures are composed of elements which are themselves structures of organized elements, where the principles that govern the organization remain the same at each level. This technique allows arbitrarily large constructs to be built by the repeated use of a limited amount of processing, and this is widely employed throughout nature. Plants and the architecture of termite nests are obvious examples. In the limit, recursive techniques have no sense of scale. Thus, it can be argued that any bend in a curve on double logarithmic graph paper would immediately define a scale. Consequently, recursive structures would be expected to give a straight-line distribution (Scarrott, 1981). However, exactly how English word usage is related to such construction techniques has not been detailed, but the implications are clear.

Information theory assumed an independence between symbols based upon the Markov chain (the ergodic process) in its measure of the information carried in a message, whereas 'meaning' is dependent upon a much broader context.

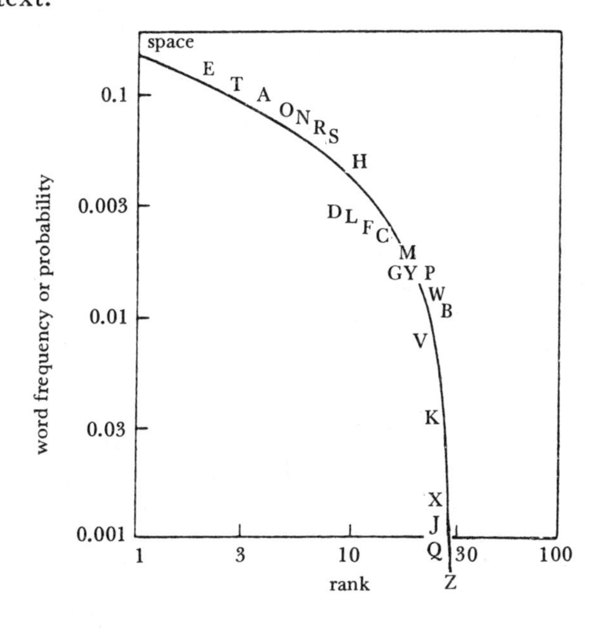

Figure 9.5 *Rank/frequency graph for English letters*

Scarrott conjectured that there is a concept of usable entropy based upon a recursive model of meaningful information that is clearly different from communicable entropy based on the Markov process. Scarrott, however, did not define the model or give examples of the operation. Nevertheless, it can be seen that usable entropy would be related to the number of distinct meaningful messages that can be constructed by a recursive process from the communicable entropy. It would be expected that this usable entropy, based upon the recursive process of construction, would increase at a much slower rate than the possible communicable entropy, simply because the context becomes more restricted as the meaningful message grows.

Activity can also be measured in terms of entropy. The important consequence of Scarrott's conjecture is that it suggests that meaningful human behaviour may be conveniently represented by a recursive structure. Such structures can be represented as hierarchical structures. A particular example is that when a person is diagnosing a fault in a machine, this view of the machine will be structured even if the machine is strictly monolithic. The machine will be considered to be made of parts, each part consisting of subparts and each level being different descriptions of the same machine. The behaviour of the individual and the way he reports his actions will reflect this view. Since the context of this behaviour is bound to a particular machine as well as a group of experts who communicate with each other, this structured view would be expected to be uniform throughout the group. If there is a change in this structured view, then the change will be gradual.

There is a continuity between the language used to describe and unite a group's activities and the activity itself (Wittgenstein, 1953). This is because there is a continuity, inasmuch as descriptions can be assigned meaning and activities have structure. The argument here, supported by the weak evidence of Zipf's law, suggests that a structured description of an activity such as the diagnostic process may be prised directly from an accumulation of observations of a group of experts engaged in the task of fault-finding (say).

GROWING TREES FROM ENTROPY

The diagnostic process is a sequence of tests and observations (symptoms) which lead to an action that cures the fault. If all the tests and observations are recorded along with the final action (fault location traces), then this would be the data from which a hierarchical structure may be formed. The structure creation program that uses the structure for diagnosis is called the Rapid Action Fault Finding Library Enquiry System (RAFFLES).

The objective is to sift through examples of successful fault location traces associated with each replaceable unit and generate an optimum fault location guide in the form of a multi-branching tree. The first node of the tree will represent the best test, where the best test divides the different replaceable units into as near to equal-sized and stable groups of replaceable units as possible. The second set of tests in the next level down the tree repeats this requirement for each divided group, and so on until the final test in a branch isolates a replaceable unit. The measure of quality for a diagnostic test requires that the test be reliable, provides an equal division between the classes and concentrates on probable faults.

Techniques of pattern recognition show that there is a single complex measure which will balance all these requirements so that the 'best' test may be chosen. This single measure is the 'reduction in entropy per unit time' brought about by applying a particular test.

The faults associated with each replaceable unit (or fault report in the case of software) can be considered as a series of independent symptoms that identify that unit. A trace (set of symptoms) is an actual example of a fault identification. Each symptom represents the results of an action where each action can have one or more possible outcomes. The effect of dividing the database of traces according to the result of any particular test is to reduce the entropy of the whole system with respect to the faults. That is, the faults become partially ordered. The best test will be the one that reduces the entropy by the greatest amount in the shortest possible time.

The measure of a test can be achieved by considering the statistics of each test when used without any constraints. Let the range of tests be $1 \leqslant i \leqslant T$, the range of possible outcomes of test i be $1 \leqslant b \leqslant n$ and the range of possible faults, or of replaceable units in the case of hardware, be $1 \leqslant k \leqslant K$ (see Figure 9.6).

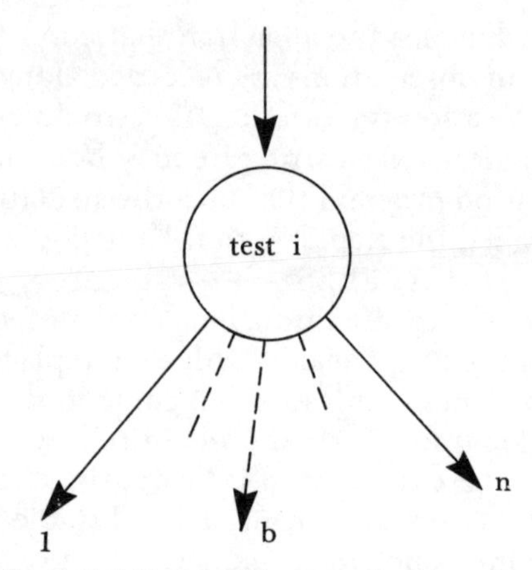

Figure 9.6 *Test i with n associated results*

We define the following probabilities:

- P(k/b, i) = probability that fault k is present (or unit k is faulty) given that test i has outcome b.
- P(k, b/i) = probability that test i will have outcome b and that fault k is present.
- P(b/i) = probability that test i will have outcome b.
- P(k/i) = probability that given test i, fault k is present.

It can be shown that, given any outcome b for test i, the entropy with respect to the set of k possible faults (or replaceable units) is:

$$H_{b,i} = -\sum_{k=1}^{K} P(k/b, i) \log_2 P(k/b, i)$$

The expected entropy $E(H_{o,i})$ of the outcome, given the test i, is:

$$E(H_{o,i}) = \sum_{b=1}^{n} P(b/i) H_{b,i}$$

This becomes, after suitable manipulation:

$$E(H_{o,i}) = -\sum_{b=1}^{n} \sum_{k=1}^{K} P(k, b/i) \log_2 P(k, b/i) + \sum_{b=1}^{n} P(b/i) \log_2 P(b/i)$$

The input entropy (ie before the results of any test are known) is:

$$H_{I,i} = -\sum_{k=1} P(k/i) \log_2 P(k/i)$$

The expected decrease in entropy $E(\Delta H_i)$ resulting from applying test i is $H_{I,i} - E(H_{o,i})$, and from the above equations this is:

$$E(\Delta H_i) = -\sum_{b=1}^{n} P(b/i) \log_2 P(b/i) + \sum_{k=1}^{K} P(k/i) \sum_{b=1}^{n} P(b/k, i) \log_2 P(b/k, i)$$

If the test i takes t_i seconds to perform, then $E(\Delta H_i)/t_i$ is the entropic decrement per unit time.

Test i is chosen such that $E(\Delta H_i)/t_i$ is maximum for the set of traces under test. This having been chosen, the next set of tests is calculated independently for each of the resulting divisions of these traces caused by applying test i.

The training set of actual test sequences that leads to the location of faults is examined, and every test is assigned its reduction of entropy per unit time. These tests can then be ordered according to this measure of excellence, and the test at the top of the list (most reduction in entropy if applied) is placed at the top of the decision tree (the first test to be applied by the user).

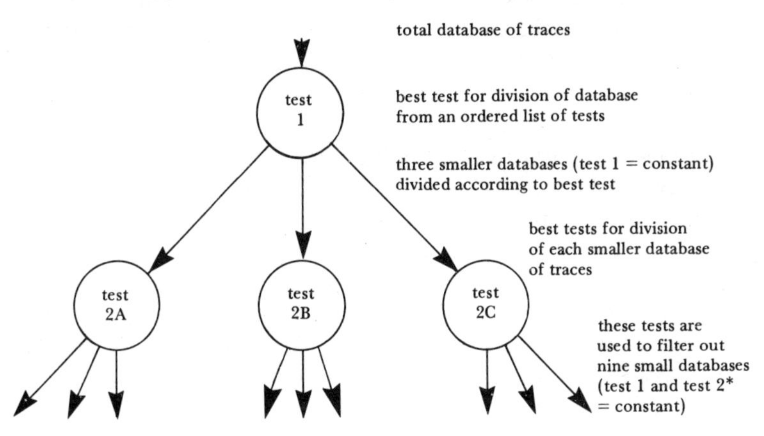

Figure 9.7 *The growing tree*

This first test, is then used to sort the training set into n smaller training sets, where n is the number of possible outcomes. The next level of the decision tree is produced by taking each of the n training sets separately and on each one repeating the initial exercise of finding the best test in the new context. In this level, the first test will be absent and the statistics in each training set will be distorted by the division. A repetition of the process will produce a collection of n best tests to be added to the n outcomes of the first test. This new larger decision tree is then used further to divide the training set. The process continues until each outcome consists of a single fault. Figure 9.7 illustrates this process.

A FOREST OF TREES

There are three useful by-products of this technique:

1. The first is that other tests which carry the same or similar information of a test, irrespective of how they are expressed, will have their measure of excellence reduced in sympathy. Thus, tests that have the same effect can be eliminated.
2. The second is that if a branch of a tree is reached where there are no more possible tests to distinguish between faults, then this will be discovered and an appropriate request can be made to the overseeing expert for a distinguishing test.
3. The third is that tests which carry little or no information can be discarded.

The method of selection of tests in creating the tree does not take into account any other factors which might influence the choice of a test. Certain tests may be too difficult for the user to perform, too trivial to be worth considering because the general fault location is obvious, or just not possible because they depend upon certain equipment. Furthermore, the engineer (or software specialist) may have additional knowledge of the local environment.

Some of these problems can be overcome by creating different kinds of trees for different purposes and by providing a mechanism for using the trees effectively with inadequate information. The existence of a test at any point in the tree is totally dependent upon its context. A test out of context has little or no meaning to the system. However, there are four possible strategies which can resolve this problem.

The first strategy is to assign each test to a classification representing the skill required to perform that test. Different trees can be created, according to the level of skill of the users.

The second strategy is to consider the 'patient' mechanism as being constructed in some hierarchical fashion. A tree can be grown which calculates the reduction in entropy with respect to a level in this hierarchy instead of the final replaceable units (or patch). Once the first level has produced a tree, further trees can be grown at the second level, one for each of the substructures. This is continued for each level until the final set of trees is formed that isolates the faults (the lowest level in the hierarchy). This is the technique of *focus imposition.* The advantage of focus imposition is that any expert user can jump directly to the appropriate level in the hierarchy, thus bypassing the intermediate stage.

The third strategy, which can be combined with the second strategy, depends upon the principle that any fault location tree generated can be described in terms of inference rules by tracing every path to a fault, thus:

IF path THEN fault

where 'path' is a list of conjoined tests and results. Such rules tend to be complex especially if several paths lead to the same fault (as different leaves of the tree). In the latter case, the conditional part of the rule will contain disjunctions. The rule can be simplified by replacing the sequence of tests and results that lead to a particular intermediate substructure in the hierarchy by its name (use of focus imposition). A single rule can then be reduced to several smaller rules, thus:

IF path 1 to intermediate substructure *THEN* intermediate substructure
IF intermediate substructure and path 2 *THEN* fault

As before, the same intermediate substructure may be reached by alternative paths, and the conditional component of the rule can become complex. The generation of these rules from these trees is referred to as *knowledge refining* or *rule induction.* The advantage of generating such rules is that it presents the 'knowledge' in a form compatible with rule-based inference systems.

The fourth strategy is required when a test given to a user is answered by 'don't know'. A problem arises in that the system then has to simultaneously progress down the tree for every possible result of a given test. If other tests are subsequently answered by 'don't know', then this will create a very large number of simultaneous paths to be traversed. The management of such a multitude of traversals becomes extremely difficult.

One solution is not to traverse the paths simultaneously, but to choose consecutively the most-likely-to-succeed path first. If a solution is found before all the paths have been traversed, the process can halt. However, every answer at a particular node must be given an order of precedence indicating the most likely path to reach a conclusion. This solution combined with rule induction becomes equivalent to applying Certainty Factors to rules.

RECURSIVE KNOWLEDGE REFINING

Once a test has been chosen by calculating its entropic decrement, then it has been assumed that every trace will be able to employ that test in order to decide in which category (branch of the tree) it should be placed. This would be the case if the structure of the activity is both hierarchical and statistically uniform. To illustrate the problem, consider a set of traces generated by following every path in a binary tree (see Figure 9.8). The synthesis process of knowledge refining described should re-create this tree from these traces. However, if the frequency of traces is modified so that the paths down the left-hand side of the tree are enhanced, then the tree re-created will be only the left-hand side because the construction of the left-hand side reduces the entropy in the quickest possible time. This tree cannot cope with any of the traces associated with the right-hand side because none of the right-hand traces provides an answer to the root test (test 2 in Figure 9.8).

The geometrical expansion of results prevents a simple 'don't know' response to each test by letting each right-hand trace filter through every path of the left-hand tree during construction. The discarded right-hand traces can now be collected, and the synthesis process can then be applied to these discards. The results will be the right-hand tree.

The two trees are linked by reclassifying all the left-hand traces so that they belong to one set and the right-hand traces to another. The process can be repeated in order to determine the best set of tests to distinguish between these two sets. This 'link' tree can be used to decide which one of the two subtrees is required. This will then reconstruct the original tree despite the strong statistical bias injected into the system. This extended process is illustrated in Figure 9.9.

It has been assumed that there will be discards only at the first application of the generate tree. In general, each application of the generate tree could potentially cause discards. It is through the recursive use of the above

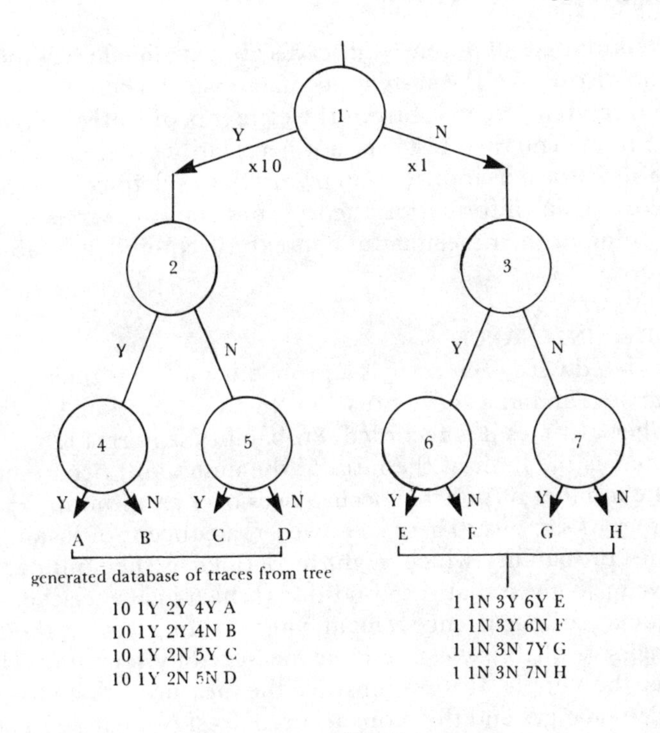

Figure 9.8 *A simple binary tree*

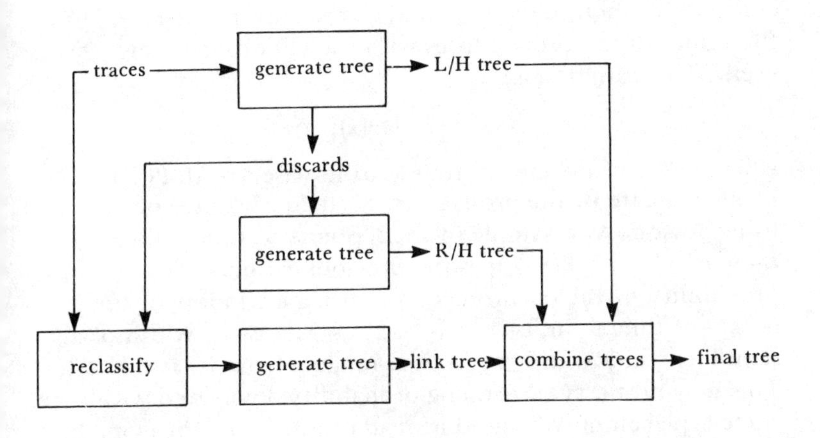

Figure 9.9 *The reapplication of the generate tree process*

procedure that all potential discards are examined until there are no discards or the system becomes cyclic. The tree is hence created piecemeal from the elements of statistically ideal trees. The final tests are not necessarily ordered according to their information properties as defined by a Markov chain (information theory), but they are ordered according to their meaningful context as required by usable entropy.

TREE MAINTENANCE

Associated with each trace is a probability of that trace occurring, and in a static situation this is proportional to the number of times it is recorded. Probability is normally defined as the limit of the ratio of the number of occurrences to the number of possible occurrences over an infinite number of samples. There is, however, a concept of instantaneous probability which might be defined as the limit of the above ratio given that the conditions which influence the frequency of occurrence remain constant. In practice, the changing probability can only be measured by sampling. The larger the sample, the less sensitive the measure will be to sudden changes and the more accurate to slow changes. The choice of sample size will depend upon the conditions.

Rather than create partitioned samples, it would be convenient to use a 'running' probability. A running probability is a function which modifies the current estimates of the values of the probabilities with each additional new event. This function is:

$$nP(x)_t = (n - 1)P(x)_{t-1} + \alpha$$

where $\alpha = 1$ if x occurs at time t, otherwise $\alpha = 0$; $P(x)t$ is the estimate of the probability of the occurrence of x, using a window or sample of the n points of time $t - n + 1$, $t - n + 2, \ldots, t$; $P(x)_{t-1}$ is the previous estimate of the probability of the occurrence of x, using a window of the n points of time $t - n, t - n + 1, \ldots, t - 1$; n is the width of a sampling window moving in time ranging from $t - n + 1$ to t. This new property of running probability associated with each trace type can now be used instead of a trace count from which the entropy is calculated.

The size of the window (n) can be adjusted according to the conditions, so that if, for example, a new variation of the mechanism under observation becomes current due to modifications, then during its initial stages in the field the window can be made small and sensitive to change.

The probability associated with each trace can be represented by:

$$P_0(f_k, s_{11}, s_{21}, \ldots, s_{ib}, \ldots, s_{TN})$$

where f_k is the kth fault and s_{ib} is the symptom resulting from applying test i and obtaining answer b. The assumption here is that the consequence of every test is known for every fault f_k. In practice, only a subset of the set $s_{11}, \ldots, s_{ib}, \ldots, s_{TN}$ is usually available.

Let S_p be a particular pattern of symptoms where every test result is known, then

$$P_0(f_k, S_{p1}) = P_0(f_k, s_{11}, s_{21}, \ldots, s_{ib}, \ldots, s_{TN})$$

where S_{p1} is a sufficient, but not a necessary, condition for f_k. If:

$$S_{ps} \subseteq S_{p1} \cup S_{p2} \cup \cdots \cup S_{pr} \cup \cdots \cup S_{pR}$$

for a selected set of S_p, then:

$$(f_k \wedge S_{ps}) \rightarrow (f_k \wedge S_{p1}) \vee (f_k \wedge S_{p2}) \vee \cdots \vee (f_k \wedge S_{pr}) \vee \cdots \vee (f_k \wedge S_{pR})$$

Thus, when the combination $f_k \wedge S_{ps}$ occurs, a possible update procedure of the probabilities $P(f_k, S_{pr})$ is to distribute an equal portion of α to each, ie:

$$nP_0(f_k, S_{pr}) = (n-1)P_0(f_k, S_{pr}) + \beta$$

where $\beta = \alpha/|(f_k, S_p)|$ ie α divided by the number of items in the set (f_k, S_p).

Although this procedure will not eliminate the accidental test results, the effect of these erroneous results in the entropic decrement calculations will be reduced as the essential test results are enhanced.

The conditional probabilities required for the entropic decrement calculations can now be restated as:

$$P_0(k/i) = \sum_{r=1}^{R} P_0(f_k, S_r/k, i) \Big/ \sum_{r=1}^{R} \sum_{k=1}^{K} P_0(f_k, S_r/i)$$

$$P_0(b/i) = \sum_{k=1}^{K} \sum_{r=1}^{R} P_0(f_k, S_r/b, i) \Big/ \sum_{k=1}^{K} \sum_{r=1}^{R} P_0(f_k, S_r/i)$$

$$P_0(b/k, i) = \sum_{r=1}^{R} P_0(f_k, S_r/k, b, i) \Big/ \sum_{r=1}^{R} P_0(f_k, S_r/k, i)$$

When the first test has been chosen and the test set sorted according to the various outcomes of this test, the associated probabilities of each trace must be modified within the new partitioning of the training set. Thus, given test i with outcome b, the new trace probability P_{ib} is:

$$P_{ib}(f_k, S_r) = P_0(f_k, S_r/i, b) \Big/ \sum_{k=1}^{K} \sum_{r=1}^{R} P_0(f_k, S_r/i, b)$$

Conclusions

The RAFFLES method of knowledge refining has been used in practice on small samples (Martin, 1984). Further evidence from other work in producing a diagnostic aid (Addis, 1980), together with some provisional results of applying this technique without recursive refining both to pattern recognition problems and to a database of fault traces of a major operating system, suggest that it can reveal a hierarchical structure from the meaningful behaviour of a group of experts. However, there is the suggestion that this technique would not be suitable for machine-generated features, such as the pattern of bits in a set of registers which are given as symptoms of a fault. As Scarrott's conjecture implies, there is no structure imposed upon this view of the machine and hence advantage cannot be taken of usable entropy.

There is another problem which arises with techniques based upon statistics. This is 'what should be done when there are usually only one or two examples of each pattern within the set'. The calculations of probability become both inaccurate and inappropriate under these conditions. Asking the expert to make some judgement about the probabilities is often fraught with danger, although the figures can often be adjusted if the results are too outrageous. It is for this reason that work using production rules and belief systems (Shortliffe, 1976) seemed to avoid these problems but they in turn have problems of consistency checking.

The generation of rules through 'induction' using the techniques of recursion and focus imposition would seem to avoid many difficulties. Consistency is checked, similar tests are detected, subjective errors are avoided and new rules are generated to account for changing patterns of behaviour. However, automatic induction that can give fresh hypotheses in the manner implied by Lakatos (1978) is not modelled by the technique of induction. This technique will not recommend new tests or suggest new approaches; it refines a description according to the principles of information theory, and it is from this theory that the rules are deduced.

One possible component of experts' activities worth investigating is the order in which they apply their tests. This should give additional information on the structure of behaviour. Currently, this information is being ignored, yet it might be used to calculate weights that modify or replace probability calculations based upon a few examples.

It would be rash to suggest that a simple hierarchical model is going to be sufficient to capture all the behaviour of a group of experts. However, the utility of this method of knowledge refining is that it provides a completely automatic system of knowledge maintenance for an expert system where there are a few thousand possible goals, and this knowledge is kept in good order for the wider group of users.

CHAPTER 10

Computer 'Understanding'

> Grammar is substantially the same in all languages even though it may vary accidentally.
>
> *Roger Bacon (1214-1294)*

Introduction

Communicating with a computer using a natural language, such as English, is considered by many to be the ideal method for linking non-computer professionals to large and possibly complex computer systems (Mylopoulos *et al.*, 1976; McCalla and Sampson, 1972). The main criticism of this approach is that the amount of computation to analyse sentences given in English is large and subject to certain ill-defined constraints. The question arises whether in the computer environment a language might avoid these complexities yet still contain the essential qualities of English that make it easy to use. For example, in ordinary speech much is often left unsaid leaving the listener to infer the meaning from the context. This ability to infer meaning is related to people's ability to increment their knowledge from conversation. Natural language has always been the best means through which knowledge is communicated (Krauss and Glucksberg, 1977). Much of the preceding work has avoided a direct confrontation with the problems of natural language analysis by insisting that the client and users conform to certain well-specified disciplines. It is important in the understanding of knowledge and its relationship with computers to appreciate the difficulties of direct interpretation of natural language by a machine; computer 'understanding' is not easily achievable without human intervention.

It is generally accepted that natural language is one of the most efficient ways knowledge can be transmitted from one person to another. In particular, it is easy to use because it employs the natural abilities of people (Lenneberg, 1967).

Natural language can be exact but more often than not it has many levels of precision for the reason that there is no virtue in showing a greater number of distinctions about the world than are necessary for the purpose in hand. It is this latter point that makes computer understanding of natural language potentially possible. For certain uses such as question answering systems, there is no need for a computer to have details of anything other than a very limited universe. Limiting the universe also limits the interpretation problem since the number of interpretations at any instance is reduced. Other uses such as machine translation between natural languages may have a much wider scope but the details of the world can be kept to a small number of generalities. However, difficulties arise in choosing the features that encompass all the distinctions required, and it is to this end that techniques, such as Extended Relational Analysis (ERA), have been devised.

Linguists are mainly concerned with the description of the generation of written text, and not all their theories have been adequate for use in the development of mechanical analysis (Lyons, 1968). As a result, many of the workers in this field have had to extensively modify linguistic theories. Language is particularly difficult to handle because of its apparent inconsistencies and illogical structure. Any rule discovered can cause people in the field to generate a number of exceptions, and any system can soon be confounded by the application of an appropriately deviant sentence. The success of a system must be measured by its ability to cope with the common run-of-the-mill utterances. Most authors of systems tend to have a set of paradigm sentences (often related to a specific task domain). To know their paradigms is often to know their systems.

Computer understanding has developed from a variety of different points of view but despite this there has been a general convergence in the way the systems work. In this study, the systems have been split into first and second generations (Simmons, 1965, 1970). The first generation systems are based on word classification and ordering and often use statistical methods in their analysis. However, communication involves more than statistical inference, and this leads on to second generation systems which introduce context dependency and semantics, dropping the statistical techniques altogether. The keystone of the semantic description depends upon limiting the universe of discourse to a few common relationships, and the link between semantics and syntax is called 'word-government'. The 'word' proves to be the label to which a set of partially formed semantic structures can be associated. These are put together according to rules of grammar to form a larger semantic structure that represents the meaning of a sentence.

There are five main reasons for work in natural language understanding systems. These are for a computer to:

1. form a database for fact retrieval from documents written for human use.
2. allow a user to make enquiries of a database in a natural and unconstrained manner;
3. accept the statements about a problem generated in a natural way and be able to interact so as to expose missing elements and inconsistencies in the statements;
4. provide a fast unbiased transformation of one natural language into another natural language in such a way that the meaning of a message remains constant (ie automatic translation);
5. model human verbal behaviour so that insight might be gained into intellectual mechanisms.

This study shows that the processing elements required for natural language understanding are substantially the same in most systems even though the details of their operation may vary.

First generation systems

The first suggested use for machine processing of natural language was automatic translation. This idea was originally conceived by Booth in New York in 1946 (see Booth *et al*, 1958). Between 1949 and 1952 several papers were written on the topic and it was not until 1954 that the first machine translation of simple Russian sentences to English sentences was made at Georgetown University. The vocabulary was restricted to about 250 words. Work on machine translation commenced in England in 1955 at Birkbeck College, University of London and in Moscow in the same year.

Research towards natural language question answering systems, on the other hand, started in 1959 and, up to the beginning of 1965, 15 or 16 programs had been written which answered some types of questions presented in English. None of them were very practical due to their limited capabilities. For example, in 1959 Green allowed a computer to carry on a seemingly intelligent conversation about the weather. In this program the problem of analysis was avoided by searching for keywords like 'rain', 'July' and 'not'. According to the implied intention of these words, a standard response was selected and keywords were inserted in the blank positions. Thus:

User: 'I do *not enjoy rain* in *July*'
Response: 'Well, we don't usually have *rainy* weather in *July* so you will probably *not* be disappointed'

The early machine translators and information scientists were misled by assuming that the 'word' is the basic unit of meaning. The paradigms were that mechanical translation required the simple discovery and substitution of target language equivalent words and that information retrieval requests and data structures can adequately be represented by some combination of words with logical connectives. These paradigms were extended by the use of a large thesaurus of words, statistical association probabilities and superficial syntactic structures. All this proved inadequate in the light of the Automated Language Processing Advisory Committee (ALPAC) (1966) report. These first generation systems were further handicapped by the lack of adequate linguistic models, the use of low-level languages such as Information Processing Language (IPL) and limited technology. The advent of conversational computing on time-sharing systems greatly eased the task of programming and debugging complicated language processing programs. It also produced a plethora of new uses for natural language understanding programs, so providing greatly increased motivation for developing them. As a consequence, after 1966 a large number of language processing systems were constructed.

Second generation systems

Up to the mid 1960s natural language processing had not extended far beyond a context-free grammar with a set of *ad hoc* rules to eliminate some ambiguities. For instance, in the simple case of the sentence 'Time flies like an arrow', where 'time' could be a noun, verb or adjective; 'flies' could be a plural noun or third person verb; 'like' could be a noun, preposition, or verb and 'arrow' could be a singular noun or adjective, it is possible to form 36 different analyses of this sentence. Apart from the fact that a large number of sentences can become indeterminate in their final structure, there is a considerable amount of intermediate processing which results in no outcome at all.

In 1966 Martin Kay of Rand Corporation published a paper describing a tabular parser for context-dependent grammars. The importance of this work is that it presented a formal method of defining a phrase-structured grammar with context dependencies. The program was capable of accepting the rules of transformational (Chomsky, 1965) grammar and working on several branches of the analysis simultaneously. Experiments with the program showed that a set of rules derived algorithmically from a transformational grammar is unlikely to be the most effective or the most revealing analytic technique.

From the point of view of machine analysis of language, Chomsky's (1965) transformational grammar given in *Aspects of the Theory of Syntax* has proved to be of little use. Further work by Joyce Friedman (1969) at Stanford University who specialized in the implementation of his work, showed that the transformations are often ambiguous when used in reverse.

Thorne *et al.* in 1968 produced a program which could assign syntactic labels to an indefinite number of words while operating with a finite dictionary. The program also makes only one pass through the sentence in its analysis. They demonstrated that there are two classes of words: an open class such as nouns, whose meaning does not have to be known in order to determine its syntactic role in a sentence, and a closed class such as prepositions which plays a vital role in defining the syntax of a sentence. He estimated that the total number of words for English in the closed class would not exceed 2000.

In the same year, Woods published a paper on a question answering system specifically designed for flight bookings. His program was designed to map English sentences onto a set of special functions such as CONNECT(X1, X2, X3) which will determine if there is a flight(s) between localities X1 and X3 via X2. The meaning of a sentence in this case is defined as a program of these specialist functions. However, the work for which Woods is best known is his augmented transition network grammar that is used in the analysis of natural language sentences. A sample of a transition network is shown in Figure 10.1.

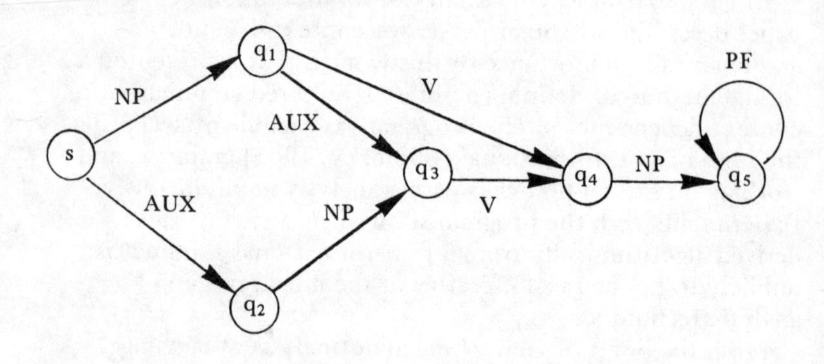

Figure 10.1 *An augmented transition network*

S represents the start state, and each arc is a test that has to be passed in order to go on to the next state. The test is usually the successful pass through another transition network such as Noun Phrase (NP). Context dependency is achieved by having special registers which preserve some facts about each state. Transitions can thus include some Boolean or heuristic algorithm using these facts.

The techniques introduced by Kay and Woods (Kay, 1966, 1967; Woods, 1968, 1970) have been widely used as a basis for natural language analysis by other workers.

The introduction of semantics

SEMANTIC STRUCTURES

A distinct theoretical advance was made for natural language processing in 1963 by the linguists Katz and Fodor. In their paper *The Structure of a Semantic Theory*, they expanded the differences between semantics and syntax. In particular, they provided a mechanism by which the semantics of a sentence could be analysed in terms of semantic features. Thus, from the two meanings of 'colourful' and the three meanings of 'ball', the unique meanings of the phrase 'the colourful ball' in the sentences:

'The boy hit the colourful ball'

and:

'They danced all night at the colourful ball'

can be clearly distinguished.

One of the first attempts to implement a program based on these concepts was made by Quillian in 1966 (His thesis was reproduced in 1968). From his ideas of the structure of a semantic memory, a program was created called the Teachable Language Comprehender (TLC) which allowed a user to insert facts as a set of simple sentences. A connective network, called a semantic net, is built of these facts which shows the relationship between different items, and this is used to analyse further sentences.

Semantic nets are graphical illustrations of knowledge structures consisting of labelled arcs and nodes. Associated with these nets are sets of procedures which allow the access, the manipulation and the construction of concepts. A semantic net also requires rules that control the development of well-formed structures. Quillian's initial work was limited to word definitions but since then many developments and extensions have been made (Ritchie and and Hanna, 1982).

There is no standard set of primitives for creating semantic nets, and every author has his own concept of what constitutes an elementary arc and node. However, the general approach is to make arcs a closed set of organizational relationships and nodes an open set of substantial (factual) relationships (or units). Organizational (structural) relationships describe a particular role of a node with respect to another node.

Phillips (1978) extended the concept of semantic nets to discourse analysis. He attempts to justify the semantic structures in terms of Linguistics, Psychology and Sociology. Phillips' primitives make greater discriminations than Quillian's primitives. For example, the relationship IST (InSTance) can indicate that 'William Proxmire' is an instance of 'person' and VAR (VARiety) can show that 'person' is a variety of 'mammal' (see Figure 10.2). Thus, VAR is a category/subcategory relationship between nodes, whereas IST is category/element relationship between nodes. TYP (TYPically) and MAN (MANifestation) link a category node with some activity that is typical or some alternative view (manifestation) of a category (or elementary object). The example given in Figure 10.2 illustrates this improvement and shows the distinction between VAR and IST. In Quillian's semantic nets, the arcs are reversed and both are labelled ISA (IS A) so that 'William Proxmire ISA person' and 'Person ISA mammal'. However, the difference between VAR and IST is needed because of the 'inheritance' of properties. Therefore, we can conclude from the net that since William Proxmire is a person and a person is a mammal which in turn is a vertebrate than 'William Proxmire was born'. On the other hand, William Proxmire might not typically sleep (because he's an insomniac) even though a 'person' typically sleeps.

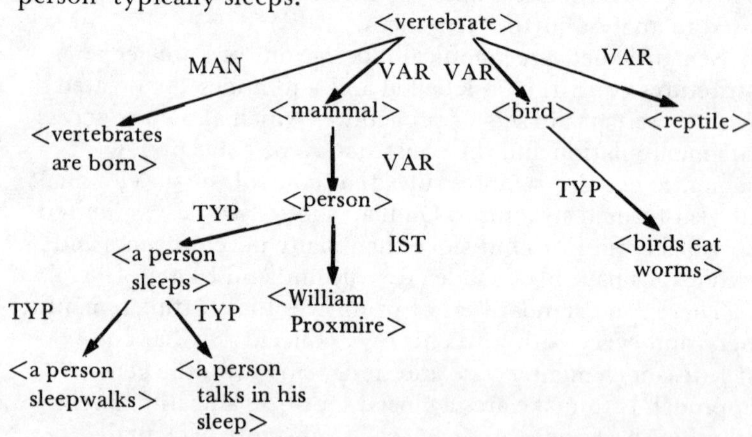

Figure 10.2 *A paradigmatic organization (structural knowledge) (Phillips, 1978)*

SUPERNODES

As with most semantic nets, Phillips relies on having different types of nodes. The adverb and adjective nodes are not distinguished since these are considered as modifiers applied through the APL arc (see Figure 10.3). P-W indicates, for example, that a 'handle' is part of the whole concept 'axe'. Of particular importance is the 'event' node introduced by Phillips, which is a means of encompassing whole propositions as a single unit. Therefore, 'Mary slapped John

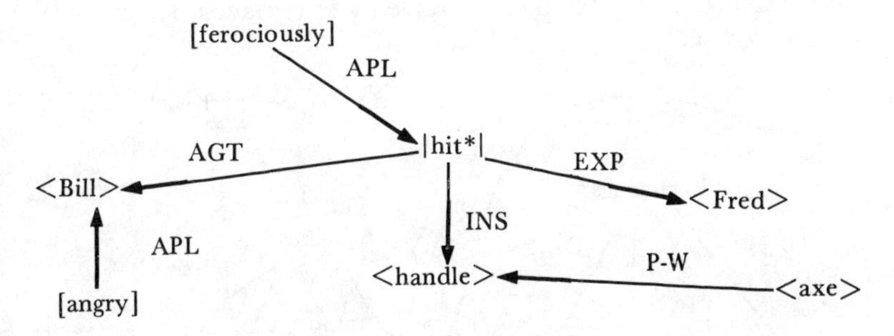

'Angry Bill ferociously hit Fred with an axe handle'

Figure 10.3 *Syntagmatic organization*

because he chased her' would be represented by the concept of 'John chased Mary' causing the event 'Mary slapped John'. This is shown in Figure 10.4.

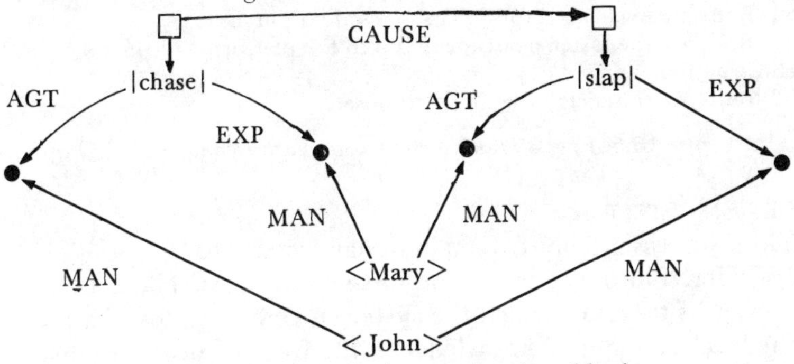

'Mary slapped John because he chased her'

Figure 10.4 *Discursive organization*

Hendrix (1975) introduced the principle of partitioning, whch can be used to the same effect (ie to define scope). Partitioning draws a boundary (space) around a part of the semantic net. This boundary can also be used as a 'focus' of attention where only the items within that space need be considered. The spaces overlap so that objects can manifest themselves in different contexts. Spaces also have labelled arcs connecting them to indicate contextual relationships (discursive relations). A simplified example of Hendrix's semantic net is given in Figure 10.5. Note that manifestation (MAN) is contained within this concept and that the normal case classifications (see p. 276) have been replaced by a specialized set.

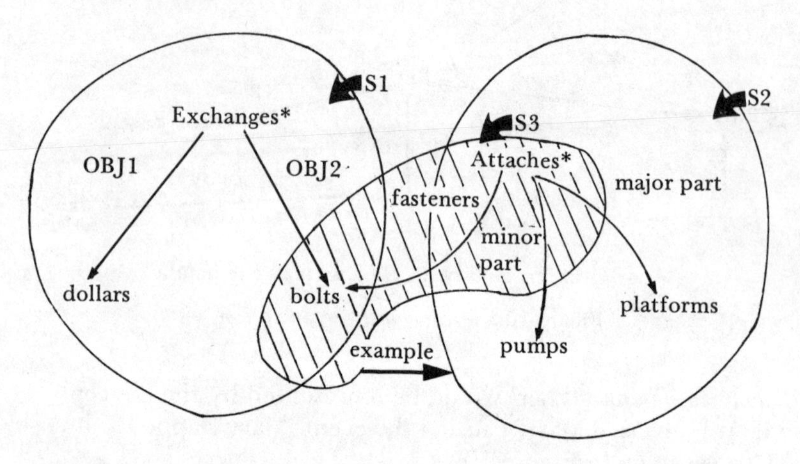

S1 'Bolts are exchanged for dollars' – context purchase
S2 'Bolts are the fasteners of the pump to the platform' – context construction
S3 'Bolts are fasteners' – context purpose

Figure 10.5 *A partitioned semantic network (Hendrix, 1975)*

DESIGN CRITERIA

There are essentially two main design criteria to be considered in the construction of semantic nets. These are the set of tasks for which the system is to be employed and the level of assumed knowledge in the form of propositions that can be expected to be known by the user. The set of tasks will suggest the kind of organizational relationships that will support the procedures needed. If the task domain is construction (ie see Figure 10.5), then the organizational primitives and the verbs with their cases will involve relative spacial relationships, methods and sequences of operations. Secondary tasks may require primitives to do with purchase, time and cost. If the task domain is diagnosis, then tests, symptoms, cause and effect may well be the dominant considerations. The choice of the technical primitives should be such that they are finite, sufficient to express and distinguish the required detail of the task, and irreducible with respect to each other in that no one set of primitives should be able to describe another (Wilks, 1977). However, other considerations may modify these ideals.

Related to task analysis is the set of appropriate procedures and their parameters. The parameters will define the form of the primitives. The procedures are needed to interpret and present the knowledge in a way that is comprehensible and useful to the user. In the construction task, for example, the terms pump and platform may need further descriptions.

It would be ideal if all descriptions required in a task domain can be accounted for by ensuring that the primitives were absolute and this knowledge was available on request. The problem of specification can be approached through first-order logic since this makes no assumptions about the world. All possibilities are open to inspection by the user. Given a particular situation the specification can be precise, providing the words chosen give adequate clues to the user as to how to map the variables and relations into the world. Once the constraints and requirements of formulating a situation are made in logic, it is possible to define the way the objects interact with each other. The methods of interpreting, manipulating and presenting this knowledge are fixed and well understood. Therefore, statements in logic can equally well be illustrated by a picture or a semantic net. Disadvantages arise from the use of logic: it is very difficult to prevent the procedures from exploring obviously incorrect solutions, also, logic is only first order, which limits the kinds of questions that can be asked and the kinds of answers available. Semantic nets (or at least the systems using that style of knowledge representation) can be very efficient as well as providing answers to higher order problems. The efficiency comes through a tighter control over the net searching procedures by employing 'heuristics'.

AN ERA OF SEMANTIC NETS

Semantic nets are directed graphs with labelled nodes, and these concepts can be expressed in terms of a semantic functional dependency (Sfd) graph. Each node is one of a fixed set of types; it can be identified (by a number) and labelled with a name. An arc is identified by two nodes and a relationship. However, an arc, representing a specific relationship, is restricted to the type of nodes it may connect. Associated with the fixed set of possible relationships is an inverse relationship. The entity sets for a first-order directed graph are:

#TYPE [Type:]	;List of allowed types of node
NODE [NodeNo: Type, Name]	;The nodes of a directed graph
ARC [NodeNo1, Rel, NodeNo2:]	;The arcs connecting the nodes
#REL [Rel:InverseRel]	;List of allowed relationships Rel and their Inverse
#TAILREL [Type, Rel:]	;List of allowed node types at tail of Rel
#HEADREL [Rel, Type:]	;List of allowed node types at head of Rel

The graph is first order because the constraints imposed in its construction are only local. The grammar of construction is fixed by the entity sets marked by a hash (#). Directed graphs of higher order may be constrained by much wider contexts than the immediate nodes and arcs. The Sfd graph given in Figure 10.6 is also a directed graph and can thus contain its own description.

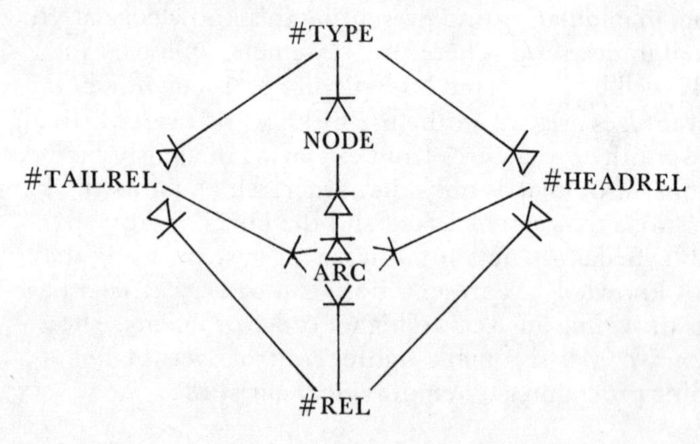

Figure 10.6 *An Sfd graph of a first-order directed graph*

Case grammar: the limited universe

Case grammar put forward by Fillmore in his paper *The Case for Case* in 1968 proved to be particularly attractive to those involved in mechanical analysis of language and the representation of knowledge. The reasons for this being that sentences could be standardized during analysis and also that it linked semantics with syntax. In case grammar, a sentence can be considered as a 'modality' plus a 'proposition':

$$S \rightarrow M + P$$

The modality defines negation, tense, mood and aspect, whereas the proposition is a tenseless structure consisting of a verb, plus noun phrases and embedded sentences:

$$P \rightarrow V + C_1 + C_2 + \ldots + C_n$$

Each C is a case name which generates either a noun phrase or an embedded sentence. At least one case must be present but no case may appear twice.

The case markers are called the Kasus functions which generate a proposition or case affix:

$$C_i \rightarrow K + NP$$

The cases related to the verb are, for example:

AGENT (A) – the instigator of the event
OBJECT (O) – the entity that moves, changes or whose existence is under consideration
INSTRUMENT (I) – the stimulus and immediate physical cause of an event

Verbs are classified according to 'case frames', which tell what case relationship may exist between a verb and its nouns. The major limitation is that there is no definitive statement as to what the case should be and those cases that do exist are not sufficiently well-defined to be unambiguous. Several papers have subsequently been written introducing new cases and redefining the old ones.

The work of Fillmore (1968, 1971) and Katz and Fodor (1963) shows two important points. The first is that there is a finite set of relations which can describe the world sufficiently to account for semantic similarity between different sentence constructions. The second is that sentence generation can be constrained to meaningful sentences by the use of word-limiting relationships (case frames).

Simmons *et al.* (1970) (who had originally written the program Protosynthese 1 based on first generation concepts) constructed a program based on the new points of view. The data structure was made up of event triples of the form (X R Y) where each X, R or Y is either a primitive element of the system, a concept corresponding to an English word sense, an instance or paraphrase of a concept or another event triple. Each entity is represented by an internally generated code number. The concepts of the system are related semantically to each other by the relation 'superset', 'equivalence', and 'subset'.

A second type of semantic relation between concepts is the semantic event form. This is a triple (X R Y) which resolves both syntactic and lexical ambiguity. The X, R and Y represent elements of the subsets of X, R and Y which can go together to form a meaningful unit. Interpretations which contain one or more disallowable combinations are rejected. The semantic event form acts in the same way as case frames.

In order to make deductions from the semantic net, each relation could have one or more of five possible properties. These are 'reflexibility', 'symmetry', 'transitivity', 'left-collapsibility'and 'right-collapsibility'. In addition, there are five relational operations that the system can use to make inferences.

The system can analyse text such as 'The stones and iron that fall to earth from outer space are called meteorites', and will respond to a question such as 'What is a meteorite?' by 'Iron that fall to earth from outer space be called meteorites. Stones that fall to earth from outer space be called meteorites'. More complex deductions than this are obtainable from the given relations in the text.

The program described by Kellogg of Systems Development Corporation in 1968 translated facts and requests for facts expressed in English into procedures that are sufficiently explicit to be directly accepted and processed by a database management system. In his paper he provides an example of how natural language can represent a very compact means of expressing operationally complex ideas. An actual request given to his program 'for the smoggy high-income cities, what is the age-income value range?' was translated into 20 separate requests to the database management system, and there was no lesser set of instructions that could be devised that would produce the same effect.

Word-government: the syntactic/semantic boundary

In the meantime, a considerable amount of work was being done on automatic indexing and abstracting. This work began in 1960. In that year a report by Maron of Rand Corporation investigated the use of computers to scan reports in order to generate automatically some kind of free indexing. It was not until 1970 that this work entered the second generation of language processing. Robison of Lockheed Palo Alto Research Laboratory in his paper *Computer-detectable Semantic Structures* suggested the use of 'word-government' as a basis for relating semantic meaning to syntactic structures.

Robison viewed human language as a vehicle for describing relationships. The parts of speech of a word are in fact binary relations. He showed that certain classes of words can determine the existence (or 'govern') other words in a sentence. Verbs, for example, can require 0, 1 or more objects. The verb 'believe' can be either transitive or intransitive and if intransitive can require the word 'in' as in 'I believe *in* all men with white ties'. It is possible to associate given meanings of certain words, such as verbs, with a sentence pattern, and it is these patterns that form word-government tables. At this point, it should be noted that these governed words include those words which have been recognized earlier as case markers. These case markers are generated by the Kasus function which provides the preposition, post-position or case affix to a noun phrase.

Lois L. Earl (1970, 1972, 1973) (RAND) extended this work of analysis of natural language for automatic indexing by using word-government. Her attitude was that it is useful and practical to try and extract as much information as possible from syntax divorced from meaning, but ultimately the syntax and semantics are inseparable, with meaning derived from syntax, and meaning determining syntax. Word-government provides an insight into this marriage between syntax and semantics, and is a way of codifying some aspects of it for use in automatic language handling. Her estimate of the complete word-government tables for English would be fewer than 8000 words. Parts of speech and exception words would require about 1000 entries.

The word-government table is used in four capacities. The first is to determine when a noun phrase must be split into two phrases to provide an indirect object as in:

1. Fred *lost* (the dog biscuits) – no indirect object for *lost*

allowed*
2. Fred *gave* (the dog biscuits)
3. Fred *gave* (the dog) (biscuits) — two semantic readings possible for *gave*

The second is to determine the word or phrase which a prepositional phrase modifies as in:

'The Supreme Court will soon hand down a judgement on the case to the Lower Court;'

can be regarded as:

'The Supreme Court will soon hand down S on S to S'

Using pattern matching against the possible allowed combinations in the word-government table (see Table 10.1), all ambiguities can be resolved. S is a noun, noun phrase or noun clause. '/' in Table 10.1 represents any string, and patterns given in parentheses are optional. Vt represents a transitive verb. The number in the meaning column is a pointer to some description.

WORD		*MEANING*	*PART OF SPEECH*	*PATTERN*
1	HAND	1	Vt	S/S
2	HAND	1	Vt	S/ to S
3	HAND	2	Vt	/ in S/(to S)
4	HAND	3, 5	Vt	/down S/(to S)
5	HAND	4	Vt	/out S/(to S)
6	HAND	5	Vt	/on S/(to S)
7	HAND	6	Vt	/over S/(to S)

Table 10.1 *Word-government table*

The third is to detect an error in apparent infinitives as in ' . . . when they were carried back to Paris . . . '. 'to Paris' might be called infinitive but the government patterns for 'carry' do not show any infinitive as a secondary. The word-government patterns do however, show:

'carry, Vt, /back S/ to S'

*Exceptions can be found by situation contrivance. For example, if Fred and the dog were partners in a competition and the prizes for dogs are biscuits, than a poor performance by Fred would mean that:

Fred lost (the dog) (biscuits)

In the fourth capacity, word-government can be used to detect an error in the resolution of a noun-verb ambiguity as in ' . . . *hold* up the liver by way of illustration . . . '. None of the noun patterns for hold includes *up*.

Word-government cannot disambiguate the meanings and relationships of nouns and adjectives within a noun phrase such as:

'engagement ring'
'engagement book'

nor can it identify the modification pattern or logical relationship within a noun phrase as in:

'new (snow suit)'

or:

'(new snow) suit'

Word-government is designed for analytical rather than generative purposes which makes it unacceptable for its integration into a theory of language.* It also does not provide a test for a meaningful sentence. Thus, 'The green idea near the tree is silent paint' is structurally sound from the word-government point of view. It can, however, differentiate between the sentences in telegrammatic style:

'Your sister suspended for minor offences'

and:

'Your sister hanged for juvenile crimes'

It is Earl's contention that eliminating the need for world-knowledge by use of word-government tables can vastly simplify the computation of the semantic knowledge without significantly reducing the power of the system for most projected uses. Her system was tested with 256 sentences from a text on the philosophy of science of which 45 sentences contained some kind of error in the final analysis.

Semantic structures: memory and meaning

Despite Earl's acknowledgement of semantics in her analysis,

---

*Linguists are mainly concerned with explaining the generation of text and psychologists with providing theories on how it is used.

her system is as pure a syntactic analysis as can be obtained while still having a practical system. There is no reference to semantic features, and this is why nonsense sentences can still be parsed. At the other extreme, Wilks in 1971 working at Stanford University produced an English-to-French translation program that was based as much on semantic analysis as is possible (Wilks, 1972, 1975a, 1975b, 1977).

The world which his semantic theory defines consists of 60 primitive elements split into five categories. These are:

1. *Entities:* Man, Stuff, Thing, Part, Folk, Act, State, Beast etc
2. *Actions:* Force, Cause, Flow, Pick, Be etc
3. *Type Indicators:* Kind, How etc
4. *Sorts:* Container, Good, Through etc
5. *Cases:* To, Source, Goal, Location, Subject, Object, In, Possessed etc

Each word in the dictionary has associated with it a formula based on these 60 primitives. Thus, a policeman would be defined as:

((FOLK SOURCE) ((((NOTGOOD MAN) OBJECT) PICK) (SUBJECT MAN)))

ie a person (MAN) who selects (PICK) bad persons (NOTGOOD MAN) out of (SOURCES) the body of people (FOLK).

The analysis is carried out with the aid of acceptable 'templates' which are Subject-Action-Object triples that represent some kind of primitive meanings. Examples of these are:

MAN – FORCE – MAN
MAN – FORCE – THING

The syntax analysis is very basic using only certain key-words such as prepositions to determine phrase boundaries. The words in the sentence are scanned, and a match is obtained between the templates and the word definitions. This will usually produce several ambiguous possibilities, which are scanned further to link up other primitive meanings which relate preferences among the words for Types etc. For example, the policeman will match with MAN in the first triplet and will ideally require a (NOTGOOD MAN) as Object. The template which produces the most preferences is the one chosen.

A further construction of the data structure which represents the phrase meaning is performed by the use of 'paraplates', which are associated with certain prepositions. The paraplates represent instructions on what to look for when relating sentence fragments analysed by the templates. Among the semantic elements being tied together are French words and phrases which represent that unit of meaning. The generation of the French is done by combining these words and phrases in the order dictated by the semantic structure.

The semantic structures created are not confined to just single sentences but extend to whole paragraphs. This allows pronoun references to be tied with objects in another sentence. Thus, the sentences:

'This is a drink.' 'Do not give it to John'

is correctly translated as:

'Ceci est une boisson. Ne *la* donnez pas à Jean'.

The art in creating these programs is that the interlingual representation should retain sufficient information for its purpose and no more.

Wilks' program, written in LISP 1.5 had a dictionary of 600 words and responded with a translation of a sentence pair within a second or two depending on the sentence structure.

In the same year, a system which caused a considerable amount of excitement in artificial intelligence circles was written by Terry Winograd at MIT (1972). His tenet is that language is a means of communicating between a speaker and listener about a shared world. The work is important because he brought together several important concepts into one complete natural language understanding system.

He was one of the first to use systemic grammar for sentence analysis (Hudson, 1973), and his work in this field has far outstretched the needs of the simple 'blocks' world for which his program (SHRDLU) was designed. This blocks world is a table top with different-sized coloured blocks and pyramids. It also contains a box in which objects can be put, an arm and a simple hand for manipulating objects. Conversations between the program and the user are limited to talking about this world. The user can issue commands to change the world such as 'Put the red block which stands on the blue one into the box' or ask questions such as 'How many red blocks are there which have a pyramid on top?' or provide new information as in 'I own all small red blocks'.

The semantics of each word are stored as a partial PLANNER theorem (Hewitt, 1969) with variables that are limited to having some property like 'physical object'. A sentence is analysed by the production of a complete theorem which has to be proved by application of it to the blocks world. The blocks world, itself, is further constrained by a set of axioms which describes the 'reality' of such objects. Thus, objects cannot be in two places at once, blocks cannot be put on top of pointed objects etc.

An intermediate position, between the detailed but limited semantics of the blocks world and the general but extensive semantics of a translation machine, has been taken by Schank and Rieger (1974). The main work concentrated on the nature of a conceptual memory which represents the semantics of a sentence and includes the inferences made by that sentence. The conceptual memory is an attempt to extend the analysis of the sentence or paragraph to include a wider frame of reference. Thus, if I say 'I am going' a very reasonable enquiry would be 'Where?'

A conceptual processor will make use of unfilled conceptual case slots to search for a given type of information in a sentence or a larger unit of discourse. The world is defined in terms of six conceptual categories shown in Table 10.2.

sign	categories
PP	real world objects
ACT	real world actions
PA	attributes of objects
AA	attributes of actions
T	time
LOC	location

Table 10.2 *Conceptual categories*

These categories relate to each other in 16 possible ways. Some are shown in Table 10.3. Thus, for the sentence:

'John hurt Mary'

the analysis would be as shown in Figure 10.7.

sign	relationship
PP ⟺ ACT	indicates that an actor acts
PP ⟺ PA	indicates that an object has a certain attribute
ACT ←O— PP	indicates the object of an action
ACT ←R— (→ PP / — PP)	indicates the recipient and the donor of an object within an action
X ⇑ Y	indicates that conceptualization X caused conceptualization Y
PP ⇐ (→ PA2 / — PA1)	indicates a state of change of an object

Table 10.3 *Category relationships*

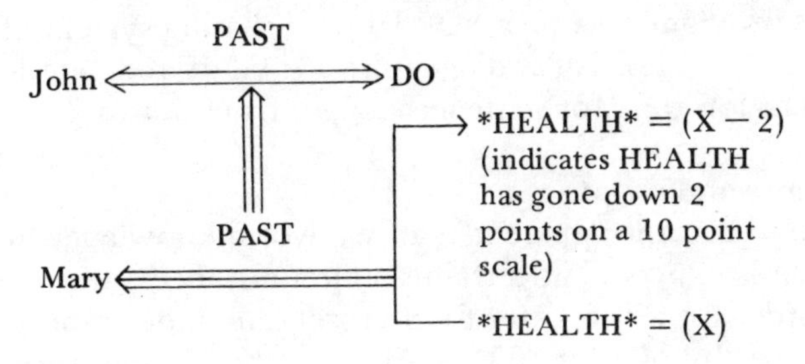

Figure 10.7 *Conceptual analysis of 'John hurt Mary'*

There are also 12 different ACTS, such as:

ATRANS – The transfer of an abstract relationship such as possession, ownership, or control

GRASP – The grasping of an object by an actor

A program has been written which uses this kind of analysis and generates inferences from it. An example of the program's response to the statement:

'John told Mary that Bill wants a book'

is

RESPONSE:

A book about what?
John possibly wants Mary to give Bill a book

Mary expected Bill to want a book
Bill wants a book
Mary knows Bill wants a book
Mary might give Bill a book
Bill thinks he would like to come to have a book
Bill thinks he would like someone to cease to have a book
Bill wants to get a book from someone
Bill thinks he would like a book to come to be near him
Bill thinks he would like a book to cease to be near someone
Bill wants to read a book

Another program based on this kind of conceptual memory was written by Goldman (1975) which produced paraphrases of input sentences. Thus:

USER:	'Othello killed Desdemona by choking Desdemona'
RESPONSE:	'Othello strangled Desdemona'
	'Othello choked Desdemona and she died because she was unable to breath'

This program runs on a PDP 10 time-sharing system, it is written in LISP 1.6 and occupies a total of 40K words. The generation time for sentences ranges from 0.5 to 2 seconds.

Frames and scripts

It soon became apparent that real world knowledge of norms (what is expected in certain set circumstances) was required in order for a computer to interpret and understand stories. A paper by Minsky 1975, known as the 'Frames paper' proposed a scheme by which standard situations are stereotyped. These stereotyped situations are called 'frames', and they guide an inference system as to what to consider and what to ignore. An example is the following story:

Jane was invited to Jack's birthday party.
She wondered if he would like a kite.
She went to her room and shook her piggy-bank.
It made no sound.

This can only be understood if the general concept of birthday parties is known; that guests should bring a gift-wrapped present, one should dress in one's best clothes and that there will be certain kinds of games and food.

An expected sequence of events for such stereotyped situations is sometimes called a 'script' rather than a frame. Going to work or school, baking a cake or buying things at the supermarket, each generally follows its own specific set of events. In general, a frame or a script is a structure that represents a collection of questions to be asked about a hypothetical situation; it specifies issues to be raised and methods to be used in dealing with them (Minsky, 1975).

An example of the use of frames is the General Understander System (GUS), which has been used in the role of a travel agent (Bobrow *et al.*, 1977). GUS converses with a client who wants to make a return trip to a city in California. The domain of discourse is restricted so that some measure of realism can be achieved without the need for vast quantities of human knowledge. It also provides the user with specific reasons for using the system, and the restriction limits the range of expectations to the user's task domain.

GUS contains several prototype frames, which serve as templates for particular instances. Most of the instances are created during the process of reasoning. A frame consists of a series of slots, each slot having a filler or a value. A frame may also have a set of associated procedures. An example of a prototype frame for dates is shown in Figure 10.8.

DATE	
MONTH	NAME
DAY	(BOUNDED-INTEGER 1-31)
YEAR	INTEGER
WEEKDAY	(MEMBER (SUNDAY, MONDAY, TUESDAY, WEDNESDAY, THURSDAY, FRIDAY, SATURDAY))

Figure 10.8 *Prototype frame for dates*

A particular date used the prototype frame as a 'model' to generate a particular instance. Thus, the instance, 28 May is given in Figure 10.9.

ISA	DATE	
	MONTH	MAY
	DAY	28

Figure 10.9 *A frame instance of 28 May*

The procedures attached to a frame are classified as either servants or demons where servants are activated only on demand and demons are activated automatically when a datum is inserted into an instance (cf optional and mandatory rules on p. 161). Frames are used here at several levels to direct the course of a conversation such that the top-level frame assumes that a conversation will follow a known pattern for making a trip. This top-level frame requires other frames to fill slots and so on until the appropriate data are collected.

Discussion

The second generation natural language understanding systems have displayed a variety of techniques and have originated from several schools of thought but, despite this, they all have several features in common.

Considering only text-analysing systems, they have all required an awareness of the morpheme (the basic unit of grammar), even though most have the word as their principal unit. Most workers have had to develop their own analytic theory of linguistics. Advances for linguistics, such as Chomsky's transformational grammar, are not always useful to the implementors. Systemic grammar and case grammar and their variations have proved to be particularly useful for two reasons. First, because systemic grammar was designed with analysis in mind, and second, because they both were concerned with the elementary use of language – to specify relationships about the world.

Another feature that is common to both systems is found at the syntactic/semantic boundary and can be isolated in its most general form as 'word-government'. The principal feature of word-government is that it relates a particular sentence pattern to a specific meaning of a word. Whether this is done directly using word-government or by constraints imposed via the mechanics of another technique is not important. Case grammar uses a limited form of word-government tied only to verbs. Winograd uses the effect of word-government with the theorems that specify meaning but this involves semantic features not found in the word-government tables. Further, the conceptual elements used by Schank also have word-government characteristics.

Wilks' templates and paraplates fall into a similar category in that they tie a meaning representation together. In his case, they are also used in reverse. For generation they are used as stereotypes producing output strings as values.

Frames, as later employed in GUS, are preconstructed semantic structures with missing elements for stereotyped situations and are a more rigid version of Wilks' templates and paraplates. Word-government is normally concerned with words within a sentence or phrase but frames indicate that constraints reach across much larger boundaries. Further, these constraints are not just associated with words but with concepts. The constraints between concepts are also expressed through the Sfd graph, and frames can be considered as structured entity sets. On the other hand, the multiplicity of rules required for word-government for any particular word can be related to its many different meanings. Thus, the underlying principle which drives word-government is essentially semantic. An example of an extended constraint was supplied by E. Charniak (1976) in his illustration of non-problem occasioned inferences (inferences that are needed to understand the story but not required to resolve word ambiguity). He considered the following text about chimpanzees, taken from the book *In the Shadow of Man* by Jane van Lawick-Goodall [(1971) reprinted here with kind permission of William Collins Sons & Co Ltd]:

> **'When Flint was very small his two elder brothers, although they sometimes stared at him, paid him little attention. Occasionally Faben, whilst *he* was grooming with his mother, very gently patted the infant, - - - - - -.'**

The reference to *he* could have been to Flint (the main subject of the paragraph) but knowledge gained much earlier in the book that baby chimpanzees do not groom confirms that *he* cannot refer to Flint, who is a baby, but to Faben. This example suggests that frames are representing the more complex and common semantic structures that can evolve recursively from simpler units (cf Scarrott's conjecture). The uncommon or special structures (which are outside the capabilities of the frame technique) must be built from the story as it progresses.

Simmons uses case grammar as his basic sentence analysis which, in turn, is being used to generate the world model in the form of a semantic net. The semantic net of Simmons, in fact, represents an extension to case grammar which makes it tend towards word-government. There are, however, no rules to include *sense*, but due to the net being built of *particular* subjects and objects, this produces a limiting criterion. Thus, 'breaking', for example, includes windows, hearts, crime rings but Simmons' net restricts the possibilities to 'John who broke a window with a hammer'.

Both semantic nets and word-government have to have special indicators that certain sentence structures have the same meaning. Hence, Simmons' dictionary will contain the tables:

LOSE:	SOURCE – X	DEFEAT:	CASUAL ACTANT – Z
	THEME – Y		THEME – X
	GOAL – Z		LOCUS – Y

where 'X lost Y to Z' versus 'Z defeated X at Y', and Earl would store, for example:

'John bought the boat from Mary.' versus 'Mary sold the boat to John.'

$$S_{1,NM} \quad \underset{Past}{Vt} \quad S_3 \quad from_{1b} \quad S_{2,NM}$$

BUY: SELL:

$$S_1^a = S_2^b$$

$$S_2^a = S_1^b$$

$$S_3^a = S_3^b$$

Further, from a generation point of view, case grammar and word-government are equivalent to transformational grammar in the sense that a given meaning requires a given sentence construction. The task of transformational grammar is to show how one sentence can be changed into another for a given variation of the meaning* or mode.

Syntax and semantic biased methods of analysis are not completely interchangeable. It is true that it is possible to interpret a sentence by using preference semantics (Wilks) and this method contains very little use of syntax, but in its ultimate form where there is no reference to word order, it is limited by the fact that it can only understand words which have a predefined semantic structure. In practice, Wilks' system does allow extensions to the language because word order is taken into account by the templates but the concern here is with the principle of preference semantics in its purest form.

*Transformational grammar is intended to provide different sentence structures without change of meaning. It is possible to extend the technique to include some changes of meaning.

Syntax, on the other hand, can specify the role a word plays in a sentence even if this word has never been used before. This technique has been used to great effect by Thorne *et al.* (1968). Hence, in the sentence 'X broke the Y', using case grammar, X is either an AGENT or INSTRUMENT and Y is DATIVE. Further conversation about X and Y can slowly build a semantic picture of them without any formal definition.

Every worker in the field of natural language will be faced with the problem, sooner or later, of constructing a complete dictionary. Earl and her co-workers were in the process of the mammoth task of producing word-government tables for all recognized governing words. In 1973 they had completed this task for all words beginning with the letters A to R, except for the letter Q. They have not mentioned in their papers that the language is changing, and their task may never really be finished. The ideal system would hopefully use the syntax/semantics characteristics to maintain local changes to the language.

All the complete systems generate a semantic and syntactic construction of the sentence or paragraph. Winograd's system forms a PLANNER theorem, Wilks' system forms a data structure with embedded French phrases and others have equivalent interlingual representations. The result is a description of the meaning that can be operated upon to perform some task, answer some question or evolve some response.

The extent to which these intermediate constructs can be used depends on the inference techniques available. The important point is that the constructs are all built from some finite set of relationships which are sufficient to perform the tasks of the system. The more detailed the requirement, such as the finite blocks world, the more relations and theorems of relational behaviour are required. In terms of effect, it does not matter if these constructs are programs to be obeyed or networks to be traced with a set of procedures. It does matter, however, if flexibility is required.

Probably, of all the five objectives given on p. 267, the most complete system is automatic translation. Principally, what is involved in translation is the construction of an interlingual representation of the meaning of a unit of discourse. This representation is built from the sections of meaning associated with the words in the dictionary and put together according to the blueprint of the syntax. The task now is to synthesize a similar structure using a dictionary and syntax of the target language. This task is actually avoided by Wilks because the structure brings back the target language phrases already embedded in the sections of meaning. For a generalized system, this is not possible. As can be seen, this total operation can be made symmetrical so that in principle any source or target language could be used given a sufficient generality of the system.

Question answering, on the other hand, requires the intermediate construction which is then used together with a knowledge base to prove or trace the existence of some criteria. Having done this, the answer must be formulated into a structure of the same kind before an answer can be generated. The flow of conversation between the man and computer should be controlled by some understanding of the purpose and behaviour (as attempted by frames) of the questioner.

The formulation of a database from documents is not really a complete system since it implies a question-answering capability and represents the first half of the task in constructing the interlingual representation of meaning. The acceptance of statements about problems is also a variation of the question-answering task.

There is still a considerable amount of work to be done in the area of semantic representation. It is still unclear as to what exactly constitutes a semantic primitive but it is certain that they are dependent upon the task domain in a very specific way. It is also indeterminate as to how much 'reasoning' (in the question-answering or inference system sense) should be expected through semantic constructs. Future work should make this semantic/'reasoning' boundary clear. It should, however, be emphasized that these classifications are only markers on a continuum ranging from morpheme analysis to intellectual insight.

Conclusions

There are currently no criteria against which to judge the success of natural language understanding systems. Each system examined is limited in what it intended to demonstrate, and all the examples provided adequate illustrations.

The real problem is that people can generate a large number of different kinds of sentences which are acceptable to others but which do not obey linguistic rules and whose form is completely unpredictable. Further, some perfectly good well-constrained meaningful sentences can be generated by people but cannot be understood by others due to the speciality of the topic, the convoluted sentence construction or the evolution of the language. Thus, it is certain that no system built will be able to cope with every possible situation.

This chapter has indicated how all the systems reviewed have tended towards the same basic paradigm consisting of four processes. These are the recognition of morphemes as basic linguistic units to act as clues to the syntactic structure; the dependency of word meaning upon the sentence structure which has been identified here as word-government and is a reflection of the underlying semantics; the generation of a semantic structure which represents some model of the task domain in the form of generalized classes and relations (this structure constrains future interpretations of associated sentences); and the operations upon these semantic structures to generate discourse about the task domain represented by these structures. Some systems are incomplete in that they may only go as far as word-government (Earl, 1972), and others have attempted to include the total paradigm (Winograd, 1972), but all the systems examined consisted fundamentally of the same elements even though the techniques and implementation have varied considerably. The main reason for this variation is that the essential knowledge used for language understanding has been divided between data structures and procedures and each system has partitioned this knowledge in different ways.

None of the systems that use semantic units to build semantic structures have given methods through which new units may be discovered. All the systems so far devised have the same essential parts. The natural language schemes used for simple applications such as information retrieval are overspecified because of these given paradigms. This latter point means that the complex definitions of 'natural' language are inappropriate for primitive uses.

Formal languages lack some vital property that is found in natural language and if this can be isolated, then a balanced, easy to use, specialized 'natural' language can be designed for different uses of man-computer interaction. This vital property of natural language is 'understanding', and understanding has been shown to depend upon a shared description of the universe of discourse between the participants. The point being that the particular form of the representation of this description is irrelevant but the *dynamics* of this representation must coincide. The representation is the image of a client's knowledge reflected in the *behaviour* of the machine.

Epilogue

Grey hairs do not make a wise man.
Old Chinese Proverb

The development of design techniques for knowledge-based systems is just beginning, and as large systems are formed more demands will be made upon knowledge engineers and designers. This will stimulate the merging of techniques drawn from software engineering, computational theory and artificial intelligence into industrially accepted methods of applied epistemics. Knowledge representation schemes will need to be specialized to cope with different task domains, and these schemes will be strongly influenced by the development of new computer architectures. Each scheme will consist of many intermediate stages (where each stage will be a controlled design step), from the client's perception of the task domain to the machine implementation of a knowledge-based system. It will produce systems that are coherent with respect to the organizations that they serve.

The particular design practice described in this book is already just one of many. There is, however, considerable similarity between the different techniques, and common principles will become clear. For example, problem-solving skills can be expressed as Petri Nets (Peterson, 1981), and Petri Nets have been forced to evolve into more complex representation schemes (Genrich and Lautenbach, 1981). These new schemes bear some similarity to the representation methods described in this book. There is the possibility that much could be learnt through finding common ground. It should be the purpose of research in knowledge engineering to find links between different descriptive and specification systems; there should be the drive to find a single unifying theory of knowledge representation.

A less machine-oriented approach to knowledge engineering will become desirable. Philosophical issues will always arise (eg Kent, 1978) even for the most primitive systems, and these issues must be faced in order to interpret and assess the qualitative component of any created system. Consequently, any training programme for knowledge engineers should include both philosophy and linguistics. Philosophy has provided many insights into the nature of knowledge, and this has resulted in a reassessment of current knowledge engineering practice. The study of linguistics is required because the nature of the knowledge-based systems will be judged by their ability to communicate.

In the preface a question was posed as to what could it mean to design a machine that understands. The machines designed according to principles that may be formally expressed show little indication of understanding. The *only* mode of reasoning available is *deduction* used within the framework of monotonic logic. This mode of reasoning has been used to 'model' induction (knowledge refining), non-monotonic reasoning (by backtracking) and abduction. Abduction (the formation of a totally new hypothesis from observations) has not been satisfactorily modelled, and the ability to detect new features of the environment to effect an alteration in perception is still a mystery. The machines mirror understanding but ultimately it remains for the client and the users to interpret a machine's behaviour.

This view of machine capability suggests to the research worker that human abilities still remain opaque. Future work in the field of knowledge engineering requires, perhaps, a change in attitude to the nature of intelligence. The self-imposed limitation that intelligence is symbol processing needs to be revised to include mechanisms that respond to the environment with internal states that are not predefined in terms of discrete symbols linked to human-biased formal methods (eg Hinton *et al.*, 1984). Such an approach was taken in cybernetics at a time when the available techniques for experimentation were primitive (Ashby, 1954), and it is time now for a reappraisal of this work in the light of modern experience.

Appendix I

The definition of application DEMOB

The DDL of the application DEMOB. This is a sub-schema of the application ODBS which is the ICL orders data base. Figure 5.9 shows the semantic functional dependency (Sfd) graph.

#APPLICATION [Provides the application name]
DEMOB

#FILE [File parameters define storage structure types. In this case, one file and differenced]

1
1

#RELATION [Specification of tables, the order of the tables defines the HEADER vector; key attributes are to the left of the colon:]

CUS [CUSTOMER: NAME]
SYS [SYSTEM:]
EQP [EQUIP: PMARK MACHINE BAR]
SAL [SALESMAN:]
CUS*SAL [CUSTOMER SALESMAN:]
CUS*SYS [CUSTOMER SYSTEM:]
CUS*EQP [CUSTOMER EQUIP:]
ORD [ORDER: AREA F550]
ORD*ITEM [ORDER ITEM: COB ODATE DISC
CUSTOMER EQUIP SYSTEM SALESMAN
TERMS SCHDATE EDDATE AMH EASV SMP CC CA]

#IMPLIES [Specification of the 0-implies between entity sets]
ORD*ITEM (ORD, CUS*SYS, CUS*EQP, CUS*SAL, SYS*EQP)
CUS*SYS (CUS, SYS)
CUS*EQP (CUS, EQP)
CUS*SAL (CUS, SAL)
SYS*EQP (SYS, EQP)

#DEPENDS [Specification of the 1-implies between entity sets]
ORD*ITEM (CUS*EQP, SYS*EQP)

```
#FIELDS         [Description of a CAFS record and data types
                eg A=ALPHA, I=INTEGER]
1 [ORDER A, 1, 0, 48) AREA (A, 1, 0, 18) F550 (X, 1, 18, 30)]
2 [CUSTOMER (A, 1, 0, 42) NAME (A, 2, 18, 72) SYSTEM
  (A, 5, 18, 36)]
2 [COB (A, 7, 6, 12) ODATE (D, 7, 18, 18) DISC (A, 8, 12, 18)
  TERMS (A, 10, 0, 18)]
3 [EQUIP (X, 2, 0, 42) MACHINE (X, 2, 0, 30) BAR (X, 3, 6, 12)]
3 [PMARK (A, 3, 18, 6) SALESMAN (A, 6, 0, 36)]
4 [SCHDATE (A, 1, 0, 24) EDDATE (A, 4, 0, 24)]
5 [ITEM A, 1, 0, 18) AMH (I, 4, 6, 24) EASY (I, 5, 6, 24)
  SMP (I, 6, 6, 24)]
5 [CC (A, 7, 6, 6, ) CA (A, 7, 12, 6)]
#OUTPUT         [gives assumed number of returned hits and style
                of output to the user]
10
TABLE (ORDER, AREA, F550)
#INDEX          [Available accessing techniques defined]
ORDER, AREA, 3
#GENERAL        [Assumed selector attribute and maximum record
                length]
ORDER
85
#MACRO          [Application dependent macro definitions
                predefined]
10
– AS AND BAR=
GB AS (CA = 9 AND CC = 9)
OH AS LIST 1 AREA F550 NAME CUSTOMER SYSTEM FOR
  ORDER=
OD AS LIST ITEM MACHINE BAR COB SMP TERMS SCHDATE
  EDDATE FOR ORDER=
EARLY AS (EDDATE SCHDATE)
LATE AS (EDDATE SCHDATE)
ONTIME AS (EDDATE LE SCHDATE)
CANCD AS (CDATE 01/01/00)
DELVD AS (DDATE 01/01/00)
VALUE AS TOTAL ALL EASV FOR ORDER
#END
STOP
```

Appendix II

BNF (MOD) definition of three states of FIDL

The following gives the syntax in modified BNF of the Flexible Interrogation and Declaration Language (FIDL) in three states. The primary state is the condition of the interpreter at the point before a conceptual model has been read from disk. The secondary state is just after a conceptual model has been read from disk, and the tertiary state is the condition where the interpreter is requesting information during updates. Other states are not included since the response is either simple and obvious or irrelevant. Figure A2.1 shows a diagram indicating the paths between the three states. The state to which a particular branch of the BNF description is relevant is indicated by subscripts. For example, ⟨HELP⟩$_2$ indicates that the branch starting with ⟨HELP⟩ is relevant only to state 2, but ⟨EDIT⟩$_{1,2}$ is available in both states 1 and 2.

Further modifications to BNF have been made in order to simplify the expression of the language. Thus, elements enclosed in [ ] mean an optional element which may be repeated indefinitely. Double obliques // mean 'in the presence of'. For example, ⟨STRING⟩//⟨NOUN⟩ means ⟨STRING⟩ in the presence of ⟨NOUN⟩. The bar | still represents the alternatives.

The elements marked with an asterisk * refer to a branch of the grammar which is only used with conceptual models of one entity set (FLAT FILES).

The ← is output by the machine as an invitation for the user to type.

⟨PARAGRAPH⟩::= ⟨SENTENCE⟩$_2$ | ⟨SENTENCE⟩$_2$; ⟨PARAGRAPH⟩$_{1,2}$ |
DEFINE$_{1,2}$ ⟨STRING⟩ AS ⟨STRING LIST⟩ |
DISPLAY$_{1,2}$ ⟨STRING⟩ |
UNDEFINE$_{1,2}$ ⟨STRING⟩ |
LINE$_{1,2}$ |
TERMINAL$_{1,2}$ |
⟨OPEN⟩$_1$ ⟨STRING LIST⟩ |
⟨OPEN⟩$_1$
⟨CLOSE⟩$_2$ ⟨STRING LIST⟩
⟨CLOSE⟩$_{2,3}$ |

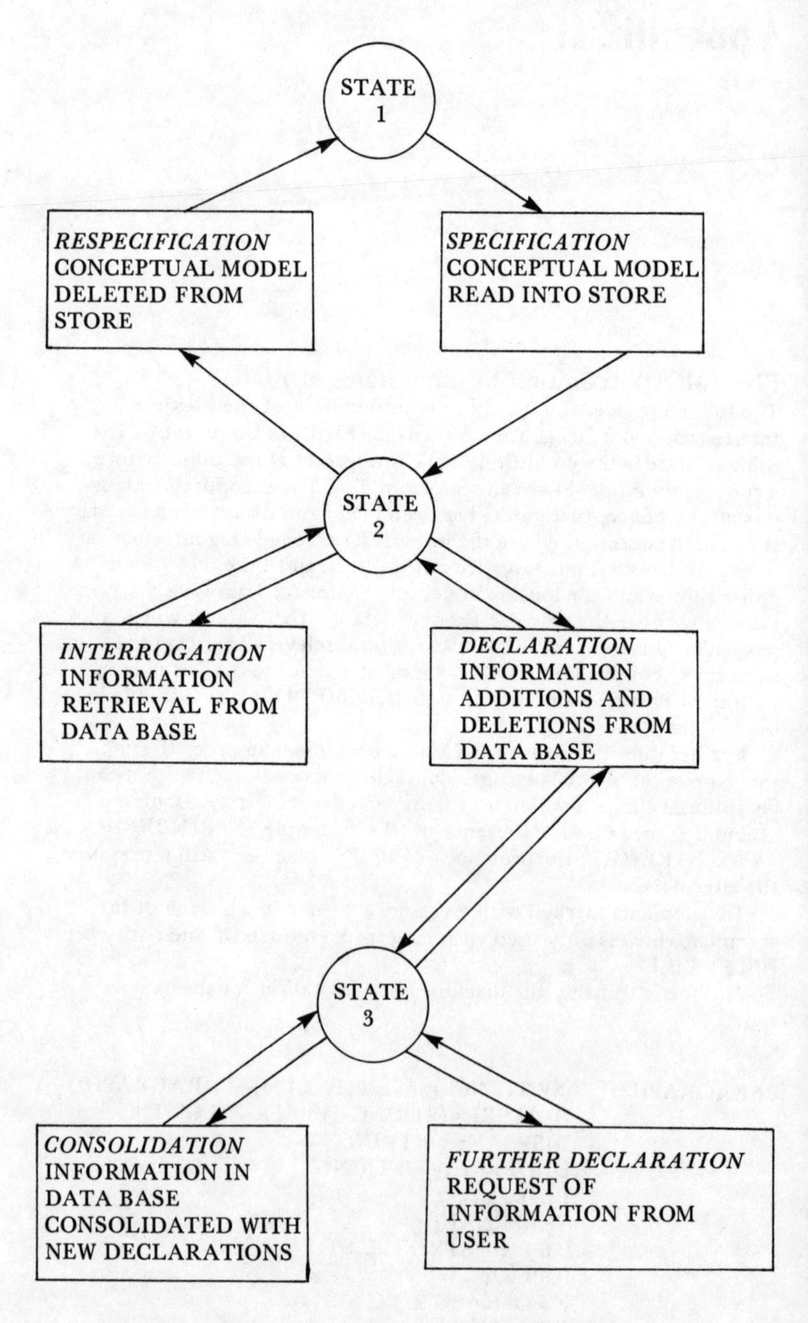

Figure A2.1. *The three main language states of FIDL*

HELP$_2$⟨STRING LIST⟩|
HELP$_2$|
EDIT$_{1,2}$|
⟨VALUE PHRASE⟩$_3$

⟨SENTENCE⟩::= FILE⟨FILE NAME⟩⟨VERB PHRASE⟩|
⟨FILE NAME⟩⟨VERB PHRASE⟩|
FILE⟨VERB PHRASE⟩
⟨VERB PHRASE⟩|

⟨VERB PHRASE⟩::= COUNT⟨COUNT PHRASE⟩|
⟨LIST⟩⟨LIST PHRASE⟩|
TOTAL⟨LIST PHRASE⟩|
AVERAGE⟨LIST PHRASE⟩|
*⟨MAKE⟩⟨MAKE PHRASE⟩|
⟨MAKE⟩⟨RETRIEVER⟩|
*⟨CHANGE⟩⟨CHANGE PHRASE⟩|
⟨DELETE⟩⟨LIST PHRASE⟩|⟨DELETE⟩⟨COUNT PHRASE⟩

⟨COUNT PHRASE⟩::= ALL⟨RETRIEVER⟩⟨FOR⟩[⟨FOR⟩]⟨SELECTOR⟩|
⟨RETRIEVER⟩⟨FOR⟩[⟨FOR⟩]⟨SELECTOR⟩|
ALL⟨FOR⟩[⟨FOR⟩]⟨SELECTOR⟩|
[⟨FOR⟩]⟨SELECTOR⟩

⟨LIST PHRASE⟩::= ⟨NUMBER⟩⟨RETRIEVER⟩⟨FOR⟩[⟨FOR⟩]⟨SELECTOR⟩|
ALL⟨RETRIEVER⟩⟨FOR⟩[⟨FOR⟩]⟨SELECTOR⟩|
⟨RETRIEVER⟩⟨FOR⟩[⟨FOR⟩]⟨SELECTOR⟩

⟨MAKE PHRASE⟩::= ⟨SUBTERM⟩|⟨SUBTERM⟩⟨LOGIC⟩⟨MAKE PHRASE⟩|
⟨SUBTERM⟩⟨MAKE PHRASE⟩

⟨CHANGE PHRASE⟩::= ALL⟨MAKE PHRASE⟩⟨FOR⟩[⟨FOR⟩]⟨SELECTOR⟩|
⟨NUMBER⟩⟨MAKE PHRASE⟩⟨FOR⟩[⟨FOR⟩]⟨SELECTOR⟩|
⟨MAKE PHRASE⟩⟨FOR⟩[⟨FOR⟩]⟨SELECTOR⟩

⟨RETRIEVER⟩::= ⟨ANOUN⟩|⟨ANOUN⟩⟨RETRIEVER⟩|
⟨ANOUN⟩⟨LOGIC⟩⟨RETRIEVER⟩

⟨SELECTOR⟩::= [⟨NOT⟩]⟨SELECTOR⟩|(⟨SELECTOR⟩)|
⟨TERM⟩|⟨TERM⟩⟨SELECTOR⟩|
⟨TERM⟩⟨LOGIC⟩⟨SELECTOR⟩|
ANY⟨NUMBER⟩FROM (⟨MAKE PHRASE⟩)
ANY⟨NUMBER⟩ (⟨MAKE PHRASE⟩)

⟨TERM⟩::= ⟨SUBTERM⟩|
⟨SUBTERM⟩[⟨NOT PHRASE⟩] TO⟨VALUE⟩

⟨SUBTERM⟩::= ⟨ANOUN⟩⟨RELATION PHRASE⟩⟨VALUE PHRASE⟩|
⟨ANOUN⟩⟨INT PHRASE⟩|⟨INT PHRASE⟩|
⟨RELATION PHRASE⟩⟨VALUE PHRASE⟩|
⟨ANOUN⟩⟨VALUE PHRASE⟩|⟨VALUE PHRASE⟩|
⟨ALPHA⟩

```
⟨RELATION
      PHRASE⟩::=   ⟨NOT PHRASE⟩⟨RELATION⟩⟨CONNECTIVE⟩|
                   ⟨NOT PHRASE⟩⟨RELATION⟩|⟨NOT PHRASE⟩
⟨INT PHRASE⟩::=    ⟨NOT PHRASE⟩⟨INTRANS⟩
⟨NOT PHRASE⟩::=    IS [⟨NOT⟩] |
                   [⟨NOT⟩]
⟨NOT⟩::=           NOT|UN
⟨VALUE
      PHRASE⟩::=   ⟨ASTRING⟩|"⟨ASTRING⟩"|'⟨ASTRING⟩'
⟨INTRANS⟩::=       PRESENT|PRES|PR|NP
⟨RELATION⟩::=      EQUAL|EQUALS|EQ|=|AS|NE|⟨|LT|LESS
                   GE|⟩|GT|GREATER|LE
⟨FOR⟩::=           FOR|FROM|WHERE|OF|WITH|WHICH|WHO|
                   WHOM
⟨LOGIC⟩::=         &|AND|OR
⟨ANOUN⟩::=         ⟨NOUN⟩|⟨TAG⟩
⟨NOUN⟩::=          Attribute Name
⟨TAG⟩::=           Relation Name
⟨CONNECTIVE⟩::=    THAN|TO
⟨FILE NAME⟩::=     ⟨STRING = 4 CHARACTERS⟩
⟨NUMBER⟩::=        Any number greater than zero
⟨ASTRING⟩::=       ⟨STRING⟩//⟨NOUN⟩|
                   1//⟨TAG⟩
                   0//⟨TAG⟩
⟨STRING
      LIST⟩::=     ⟨STRING⟩|⟨STRING⟩⟨STRING LIST⟩
⟨STRING⟩::=        Any alphanumeric character string including
                   some symbols.
                   This is a 'word'.
⟨LIST⟩::=          LIST|FIND|GET
⟨MAKE⟩::=          MAKE|CREATE
⟨CHANGE⟩::=        CHANGE|SET|ASSIGN
⟨DELETE⟩::=        DELETE|ERASE|SCRATCH
⟨OPEN⟩::=          OPEN|START
⟨CLOSE⟩::=         CLOSE|STOP|HALT|END
```

References and Bibliography

Addis, T.R. (1973) Group purchasing information system. *ICL (RADC) Technical Report* 1103, Section 4, August.

Addis, T.R. (1975a) An introduction to relational analysis for data base design. *ICL (RADC) Working Note* UG21.

Addis, T.R. (1975b) The CAFS language interpreter (FLIN). *ICL(RADC) Technical Note* TN75/1, October.

Addis, T.R. (1976) Interactive language and a relational data-base. Presented at the *Implementing Relational Data-Base Systems Symposium* University of Southampton, England, March 28-29.

Addis, T.R. (1977) Machine understanding of natural language. *International Journal of Man-Machine Studies* 9, pp. 207-222.

Addis, T.R. (1978) A feasibility study of RAFFLES. *ICL (RADC) Technical Note* TN78/3 August.

Addis, T.R. (1980) Towards an 'expert' diagnostic system. *ICL Technical Journal* May, pp. 79-105.

Addis, T.R. (1981) Applied epistemics: an empirical view of knowledge communication. *Man-Computer Studies Group* (CS/MCSG2), Brunel University,, Uxbridge.

Addis, T.R. (1982a) A relation based language interpreter for a content addressable file store. *ACM Transactions on Database Systems* 7, pp. 125-163; *ICL (RADC) Technical Report* 1209, August, 1979.

Addis, T.R. (1982b) Knowledge refining for a diagnostic aid (an example of applied epistemics). *International Journal of Man-Machine Studies* 17, pp. 151-164.

Addis, T.R. (1982c) Relational analysis for artificial intelligence. *Man-Computer Studies Group* (CS/MCSG8), Brunel University, Uxbridge.

Addis, T.R. (1982d) Expert systems: an evolution in information retrieval. *Information Technology: Research and Development* 1 (4), October, pp. 301-324.

Addis, T.R. (1983) Dependency Analysis for Relational Databases. *DBMSs – a Technical Comparison* Infotech State of the Art Report 11:5 (P.J.H. King, ed.), pp. 3-20. Pergamon Infotech Ltd, Maidenhead.

Addis, T.R.; Hartley, R.T. (1979) A fault-finding aid using a content-addressable file store. *ICL (RADC) Technical Note* TN 79/3, June.

Addis, T.R.; Johnson, L. (1983) Knowledge for Machines. *The Fifth Generation Computer Project* Infotech State of The Art Report 11:1 (G.G. Scarrott, ed.), pp. 17-30. Pergamon Infotech Ltd, Maidenhead.

ALPAC (Automated Language Processing Advisory Committee) (1966) Language and Machines – Computers in Translation and Linguistics. National Academy of Sciences, Washington, D.C.

Ashby, R.W. (1954) *Design for a Brain* Chapman and Hall, London.

Ashby, R.W. (1956) *An Introduction to Cybernetics* Methuen, London.

Austin, J.L. (1946) Other minds. In *Proceedings of the Aristotelian Society* Supplementary Volume XX. Reprinted in *Philosophical Papers* 1970, pp. 76-116. Oxford Paperbacks, Oxford.

Babb, E. (1979) Implementing a relational database by means of specialized hardware. *ACM Transactions on Database Systems* 4(1), pp. 1-29.

Babb, E (1982) Joined normal form: a storage encoding for relational databases. *ACM Transactions on Database Systems* 7(4), pp. 588-614.

Bandler, W.; Kohout, L.J. (1980) Fuzzy relational products as a tool for analysis and synthesis of the behaviour of complex natural and artificial systems. *Fuzzy Sets: Theory and Applications to Policy Analysis and Information Systems* (P.P. Wang and S.K. Chang, eds), pp. 341-367. Plenum Press, New York.

Bandler, W.; Kohout, L.J. (1985) Mathematical Relations. In *International Encyclopeadia of Systems and Control* (M.G. Singh *et al.*, eds) (in 6 volumes), Pergamon Press, Oxford and New York (in press).

Banerjee, J.; Hsiao, D.K.; Baum, R.I. (1978) Concepts and capabilities of a database computer. *ACM Transactions on Database Systems* 3(4), pp. 347-384.

Belady, L.A.; Lehman, M.M. (1976) A model of large program development. *IBM Systems Journal* 3, pp. 225-252.

Bennett, J.M. (1975) Storage design for information retrieval: Scarrott's conjecture and Zipf's law. In *International Computing Symposium* pp. 233-237. North-Holland, Amsterdam and New York.

Bennet, J.S.; Buchanan, B.G.; Cohen, P.R.; Fisher, F. (1979) Applications-oriented AI research: science and mathematics. *Stanford Heuristic Programming Project* (Memo HPP-79-22), August.

Bobrow, D.G.; Kaplan, R.M.; Kay, M.; Norman, D.A.; Thompson, H.; Winograd, T. (1977) Gus, a frame-driven dialog system. *Artificial Intelligence* **8**, pp. 155-173.

Booth, A.D.; Brandwood, L.; Cleave, J.P. (1958) *Mechanical Resolution of Linguistic Problems* Butterworth Scientific Publications, London.

Buchanan, B.G.; Feigenbaum, E.A. (1978) DENDRAL and Meta-DENDRAL: their applications dimension. *Journal of Artificial Intelligence* **11**, pp. 5-24.

Bundy, A. (1981/2) What is the well dressed AI educator wearing now? *Artificial Intelligence and Simulation of Behaviour Quarterly* Winter/Spring, No. 42-43, pp. 21-22.

Cendrowska, J.; Bramer, M.A. (1982) A rational reconstruction of the MYCIN consultation system. *The Open University* report, Mathematics Faculty, November.

Chapanis, A. (1975) Interactive human communication. *Scientific American* March, pp. 36-42.

Charniak, E. (1976) Inference and Knowledge – Part 1. In *Computational Semantics* (E. Charniak and Y. Wilks, eds), pp. 1-21. North-Holland, Amsterdam and New York.

Chen, P.P. (1976) The entity-relationship model – toward a unified view of data. *ACM Transactions on Database Systems* **1**, (1), March, pp. 9-36.

Chomsky, N. (1965) *Aspects of the Theory of Syntax*. MIT Press, Cambridge, Massachusetts.

Church, A. (1936) An unsolvable problem of number theory. *American Journal of Mathematics* **58**, pp. 345-363.

Clark, D.W.; Green, C.C. (1977) An empirical study of list structure in LISP. *Communications of the Association for Computing Machinery* **20**, pp. 78-87.

Codd, E.F. (1970) A relational model of data for large shared data banks. *Communications of the Association for Computing Machinery* **13**(6), June, pp. 377-387.

Codd, E.F. (1971a) Normalised database structure: a brief tutorial. In *Proceedings 1971 ACM SIGFIDET Workshop on Data Description and Access* San Diego, California, November 11-12, ACM, New York, pp. 1-17.

Codd, E.F. (1971b) A database sublanguage founded on the relational calculus. In *Proceedings 1971 ACM SIGFIDET Workshop on Data Description and Access* San Diego, California, November 11-12, ACM, New York, pp. 35-68.

Codd, E.F. (1971c) Further normalization of the database relational model. *IBM Research Report* **909**, IBM Thomas J. Watson Research Center, Yorktown Heights, New York, August 31. Also to be published in *Courant Computer Science Symposia* **6**: Database Systems, Prentice-Hall, Englewood Cliffs, New Jersey.

Codd, E.F. (1971d) Relational completeness of database sublanguages. *IBM Research Report* **987**, IBM, Thomas J. Watson Research Center, Yorktown Heights, New York. Also to be published in *Courant Computer Science Symposia* **6**: Database Systems, Prentice-Hall, Englewood Cliffs, New Jersey.

Coulouris, G.F.; Evans, J.M.; Mitchell, R.W. (1972) Towards content addressing in databases. *Computer Journal* **15**(2), pp. 377-387.

Date, C.J. (1981) *An Introduction to Database Systems* (3rd edition). Addison-Wesley, London.

Davis, R.; Buchanan, B.G. (1977) Meta-level knowledge: overview and application. *IJCAI* pp. 920-928.

de Dombal, F.T.; Leaper, D.J.; Horrocks, J.C. (1974) Human and computer-aided diagnosis of abdominal pain: further report with emphasis on performance of clinicians. *British Medical Journal* **1**, pp. 376-380.

Dreyfus, H.L. (1972) *What Computers Can't Do (A Critique of Artificial Reason).* Harper and Row, New York.

Duda, R.O.; Hart, P.E.; Nilsson, N.J. (1976) Subjective Bayesian methods for rule-based inference systems. *AFIPS* **45**, pp. 1075-1082.

Duda, R.; Gaschnig, J.; Hart, P. (1980) Model design in the PROSPECTOR consultant system for mineral exploration. In *Expert Systems in the Micro-Electronic Age* (D. Michie, ed.), pp. 153-167. Edinburgh University Press, Edinburgh.

Earl, L.L. (1970) Experiments in automatic extracting and indexing. *Information Storage and Retrieval* **6**, pp. 313-334.

Earl, L.L. (1972) The resolution of syntactic ambiguity in automatic language processing. *Information Storage and Retrieval* **8**, pp. 277-308.

Earl, L.L. (1973) Use of word government in resolving syntactic and semantic ambiguities. *Information Storage and Retrieval* **9**, pp. 639-664.

Edwards, W. (1972) N=1: Diagnosis in unique cases. In *Computer Diagnosis and Diagnostic Methods* (J.A. Jacques, ed.), pp. 139-151. Charles C. Thomas, Springfield, Illinois.

Fagin, R. (1977) Functional dependencies in a relational database and propositional logic. *IBM Journal of Research and Development* November, pp. 534-544.

Fairthorne, R.A. (1969) Progress in documentation. *Journal of Documentation, Machinery* **25**, pp. 330-343.

Feigenbaum, E.A. (1980) Themes and case studies of knowledge engineering. In *Expert Systems in the Micro-electronic Age* (D. Michie, ed.), pp. 3-25. Edinburgh University Press, Edinburgh.

Fillmore, C.J. (1968) The case for Case. In *Universals on Linguistic Theory* (E. Bach and R.T. Harms, eds) pp. 1-88. Holt, Rinehart and Winston, New York.

Fillmore, C.J. (1971) Verbs of judging: an exercise in semantic description. In *Studies in Linguistic Semantics* (C.J. Fillmore and D.T. Langendoen, eds), pp. 273-289. Academic Press, London and New York.

Forrester, J.W. (1975) *Collected Papers of Jay W. Forrester* Wright.

Fox, J. (1981) Statistical and non-statistical inference in medical diagnosis. February, draft document.

Fox, J.; Barber, D.; Bardham, K.D. (1980) Alternative to Bayes? A quantitative comparison with rule-based diagnostic inference. *Methods of Information in Medicine* **19**, pp. 210-215.

Friedman, J. (1969) A computer system for transformational grammar. *Communications of the ACM* **12**, pp. 341-348.

Fu, K.S. (1958) *Sequential Methods in Pattern Recognition and Machine Learning* Academic Press, London and New York.

Genrich, H.J.; Lautenbach, K. (1981) System modelling with high-level Petri Nets. *Theoretical Computer Science* **13**, pp. 109-136. North-Holland, Amsterdam and New York.

Goguen, J.A. (1976) *Robust Programming Languages and the Principle of Maximal Meaningfulness* The Milwaukee Symposium on Automatic Computing and Control.

Goldman, N.M. (1975) Sentence paraphrasing from a conceptual base. *Communications of the ACM* **18**, (2).

Gough, P.B. (1966) The verification of sentences: effects of delay of evidence and sentence length. *Journal of Verbal Learning and Verbal Behaviour* **5**, pp. 492-496.

Grant, T.J.; Elleby, P. (1984) *An AI Aid for Scheduling Repair Jobs*. Paris, Conference of the Association Française d'Intelligence et des Systems de Simulation. AI and Productivity, November 20-22.

Green, L.E.S. (1959) Conversation with a computer. *Computers and Automation* **8**, pp. 9-11.

Greenwood, R.M. (1972) Statistical distributions of diagnostic phrases of morbidity and mortality. *Bio-Medical Computing* **3**, pp. 43-57.

Hall, P.A.E. (1976) Optimization of single expressions in a relational database system. *IBM Journal of Research and Development* May.

Harrison, B. (1979) In *Introduction to the Philosophy of Language* Macmillan Press, London.

Hartley, R.T. (1982) The competent computer. *Man-Computer Studies Group* MCSG6, Brunel University, Uxbridge.

Hartley, R.T. (1984) CRIB: Computer Fault-finding through Knowledge Engineering. *IEEE Transactions on Computers* March, pp. 76-83

Hayes, P.J. (1979) The Naïve Physics Manifesto. In *Expert Systems in the Micro-Electronic Age* (D. Michie, ed.), pp. 242-270. Edinburgh University Press, Edinburgh.

Hendrix, G.G. (1975) *Partitioned Networks for Mathematical Modelling of Natural Language Semantics* Technical Report NL-28, Department of Computer Science, University of Texas.

Hewitt, C. (1969) PLANNER: A language for proving theorems in robots. In *Proceedings of the International Joint Conference on Artificial Intelligence* Bedford, Massachusetts, pp. 295-301.

Hinton, G.E.; Sejnowski, T.J.; Ackley, D.H. (1984) *Boltzman Machines: Constraint Satisfaction Networks That Learn* Technical Report CMU-CS-84-119, May. Carnegie-Mellon University, Pittsburgh, Pennsylvania.

Hodges, W. (1981) *Logic* Pelican Books, London.

Hubel, D.H.; Wiesel, T.N. (1962) Receptive fields, binocular interaction and functional architecture in the cat's visual cortex. *Journal of Physiology* **160**, pp. 106-154.

Hudson, R.A. (1973) *English Complex Sentences* North-Holland Linguistic Series. North-Holland, Amsterdam and New York.

Jackson, M.A. (1983) *System Development* Prentice-Hall International, New York and London.

Jaques, E.; Gibson, R.O.; Isaac, D.J. (1978) *Levels of Abstraction and Human Action* Heinemann Educational Books, London.

Johnson, L. (1983) Epistemics and the frame conception of knowledge. *Kybernetics* **12**, pp. 177-181.

Johnson, L. (1985) The need for competence models in the design of expert consultant systems. *International Journal of Systems Research and Information Science* **1**(1).

Johnson, L.; Hartley, R.T. (1982) A short course in epistomology and knowledge engineering. *Man-Computer Studies Group* MCSG13, Brunel University.

Johnson, L.; Keravnou, E.T. (1985) *Expert Systems Technology: A Guide* Abacus Press, Tunbridge Wells.

Katz, J.J.; Fodor, J.A. (1963) The structure of semantic theory. *Language* **39**, pp. 170-210. Reprinted in *The Structure of Language* (J.A. Fodor and J.J. Katz, eds), pp. 479-518 Prentice-Hall, Englewood Cliffs, New Jersey.

Kay, M. (1966) *The Tabular Parser: A Parsing Program for Phrase Structure and Dependency* Rand Corporation Memorandum RM-4933-PR.

Kay, M. (1967) *Experiments with a Powerful Parser* Rand Corporation Memorandum RM-5452-PR.

Kellogg, C.H. (1968) A natural language compiler for on-line data management. *Fall Joint Computer Conference* pp. 473-492.

Kelly, K.J.; Chapanis, A. (1977) Limited vocabulary natural language dialogue. *International Journal of Man-Machine Studies* **9** pp. 479-501.

Kent, W. (1978) *Data and Reality* North-Holland, Amsterdam and New York.

Keravnou, E.T. (1982) Information retrieval with fuzzy logic extension. Project report for the degree of B. Tech. Brunel University, Uxbridge.

Kohout, L.J. (1985) *A Perspective on Intelligent Systems* Abacus Press, Tunbridge Wells (in press).

Kohout, L.J.; Bandler, W. (1985) Relational product architecture for information processing. Invited paper for Special Issue of *Information Science* (in press).

Kohout, L.J.; Bandler, W. (eds) (1985) *Knowledge Representation in Medicine and Clinical Behavioural Sciences* Abacus Press, Tunbridge Wells.

Kohout, L.J.; Keravnou, E.T.; Bandler, W. (1983) Automatic documentary information retrieval by means of fuzzy relational products. In *Fuzzy Sets in Decision Analysis* (B.R. Gaines, L.A. Zadeh and A.-J. Zimmermann, eds), pp. 383-404. North-Holland/Elsevier, Amsterdam and New York.

Kohout, L.J.; Bandler, W.; Anderson, J.; Trayner, C. (1985) CLINAID: A Knowledge-based decision support system for use in medicine. In *Computer Models for Decision Making* (G. Mitra, ed.). North-Holland, Amsterdam and New York.

Kowalski, R. (1979) *Logic for Problem Solving* Elsevier/North-Holland, Amsterdam and New York.

Krauss, R.M.; Glucksberg, S. (1977) Social and nonsocial speech. *Scientific American* February, pp. 100-105.

Lakatos, I. (1978) *The Methodology of Scientific Research Programmes* (J. Worrall and G. Currie, eds) Cambridge University Press, Cambridge.

Lefkovitz, D. (1969) *File Structures for On-Line Systems* Spartan Books, New York.

Lehman, M.M. (1980) Programs, Programming and the Software Life Cycle, Imperial College of Science and Technology, Report No 80/6, 15 April. MML-259, also *Proceedings of the Institution of Electrical Engineers* **68**(9), pp. 1060-1076.

Lenat, D.B. (1977) The ubiquity of discovery. *Journal of Artificial Intelligence* **9**, pp. 257-286.

Lenneberg, E.H. (1967) *Biological Foundations of Language* John Wiley, New York.

Lyons, J. (1968) *Introduction to Theoretical Linguistics* Cambridge University Press, Cambridge.

McCalla, G.I.; Sampson, R.J. (1972) MUSE: a model to understand simple English. *Communications of the ACM* **15**(1).

McCarthy, J.; Abrahams, P.W.; Edwards, D.J.; Hart, T.P.; Levin, M.I. (1965) *LISP 1.5 Programmer's Manual* MIT Press, Cambridge, Massachusetts.

McCleary, R.A.; Moore, R.Y. (1965) *Subcortical Mechanisms of Behavior* p. 32. Basic Books, New York.

McCorduck, P. (1979) *Machines Who Think*. W.H. Freeman, San Francisco.

Maier, D. (1983) *The Theory of Relational Databases* Pitman, London.

Maller, V.A.J. (1979a) A content addressable file store. Presented at the *IEEE Spring Computer Conference* San Francisco.

Maller, V.A.J. (1979b) The content addressable file store CAFS. *ICL Technical Journal* 1, (3), pp. 265-279.

Mamdani, E.H.; Efstathiou, J. (1984) An analysis of formal logics as inference mechanisms in expert systems. *International Journal of Man-Machine Studies* 21(3), pp. 213-227.

Mandlebrot, B. (1977) *Fractals, Form, Chance and Dimension* W.H. Freeman, San Francisco.

Maron, M.E. (1960) *Automatic Indexing: An Experimental Inquiry* Rand Corporation Report, RM-2601.

Martin, M.V. (1984) Evaluation of RAFFLES and the construction of an Expert System based upon a design for a second order RAFFLES. Final year project for BSc Computer Science. Brunel University, Uxbridge.

Meredith, P. (1966) *Instruments of Communication: An Essay on Scientific Writing.* Pergamon Press, Oxford and New York.

Michie, D. (1979) Problems of the 'human window'. Presented at the AISB Summer School on Expert Systems in Edinburgh, July. In *Expert Systems in the Micro-electronic Age* Edinburgh University Press, Edinburgh.

Minsky, M. (1975) A framework for representing knowledge. *The Psychology of Computer Vision* (P. Winston, ed.), pp. 211-277. McGraw-Hill, London and New York.

Mitchell, R.W. (1976) Content addressable file store. Presented at the *On-Line Data Base Technology Conference* April.

Mylopoulos, J.; Borgida, A.; Cohen, P; Roussopoulos, N.; Tsotos, J.; Wong, H. (1976) Torus: a step towards bridging the gap between data bases and the casual user. *Information Systems* Volume 2, pp. 49-64. Pergamon Press, New York and Oxford.

Newell, A. (1980) The Knowledge Level. *Artificial Intelligence Magazine* 2(2). Presidential Address, AAAI Stanford University, August.

Newell, A.; Simon, H.A. (1963) GPS, a program that simulates human thought. In *Computers and Thought* (E.A. Feigenbaum and J. Feldman, eds), pp. 279-293. McGraw-Hill, London and New York.

Newell, A.; Shaw, J.C.; Simon, H.A. (1963) Empirical explorations with the logic theory machine: a case study in heuristics. In *Computers and Thought* (E.A. Feigenbaum and J. Feldman, eds), pp. 109-133. McGraw-Hill, London and New York.

Nilsson, N.J. (1980) *Principles of Artificial Intelligence* Tioga Publishing Co., Palo Alto, California.

Nordyke, R.A.; Kulikowski, C.A.; Kulikowski, C.W. (1971) A comparison of methods for the automated diagnosis of thyroid dysfunction. *Computer Biomedical Research* 4, pp. 374-389.

Oddy, R.N. (1977) Information retrieval through man-machine dialogue. *Journal of Documentation* 33, pp. 1-14.

Palermo, F.P. (1972) A data base search problem. *IBM San Jose Research Report* RJ1072 IBM Thomas J. Watson Research Center, Yorktown Heights, New York.

Pareto, V. (1897) *Cours d'Economic politique* Volume 2, Section 3, Lausanne.

Pauker, S.; Gorry, G.A.; Kaffirer, J.P. (1976) Towards the simulation of clinical cognition – taking a present illness by computer. *American Journal of Medicine* 60, pp. 981-996.

Peterson, J.L. (1981) *Petri Net Theory and the Modelling of Systems* Prentice-Hall, Englewood Cliffs, New Jersey.

Phillips, B. (1978) *A Model for Knowledge and its Application to Discourse Analysis* KSL-9, June, University of Illinois, Department of Information Engineering.

Pople, H.E. (1977) The formation of composite hypotheses in diagnostic problem solving – an exercise in synthetic reasoning. *International Joint Conference on Artificial Intelligence* 5, pp. 1030-1037.

Quillian, M.R. (1968) *Semantic Memory, Semantic Information Processing* (M. Minsky, ed.), pp. 227-270. MIT Press, Boston, Massachusetts.

Quillian, M.R. (1969) The teachable language comprehender: a simulation program and theory of language. *Communications of the ACM* **12**, pp. 439-476.

Quinlan, J.R. (1979) Induction over large data bases. *Stanford Heuristic Programming Project* Memo HPP-79-14 (Report No. STAN-CS-79-739).

Ritchie, G.D.; Hanna, F.K. (1982) Semantic Networks – a general definition and a survey. Electronic Lab Report, University of Kent, also (1983) *Information Technology: Research and Development* **2**, pp. 187-231.

Robison, H.R. (1970) Computer-detectable semantic structures. *Information Storage and Retrieval* **6**, pp. 273-288.

Rogers, W. (1969). *Think* Weidenfeld and Nicolson, London.

Sacerdoti, E.D. (1974) Planning in a hierarchy of abstraction spaces. *Artificial Intelligence* **5**, pp. 115-135.

Scarrott, G.G. (1973) Personal communication.

Scarrott, G.G. (1981) Some consequences of recursion in human affairs. *IEE Proceedings* **129A**(1), January, pp. 66-75.

Schank, R.C.; Rieger, C.J. (1974) Inference and the computer understanding of natural language. *Artificial Intelligence* **5**, pp. 373-412.

Shaket, E. (1976) Fuzzy semantics for a natural-like language defined over a world of blocks. *Artificial Intelligence* Memo 4 (University of California at Los Angeles).

Shannon, C.E.; Weaver, W. (1949) *The Mathematical Theory of Communication* University of Illinois Press, Urbana, Illinois. (published 1964).

Sharman, G.C.H. (1975) A new model of relational data base and high level languages. *IBM Technical Report* **12**, p. 136.

Sharman, G.C.H. (1976) A constructive definition of third normal form. In *Proceedings of ACM SIGMOD Conference on Information Retrieval of Management Data* Washington, D.C.

Shortliffe, E.H. (1976) *Computer-Based Medical Consultations: MYCIN* Elsevier, New York.

Simmons, R.F. (1965) Answering English question by computer: a survey. *Communications of the ACM* **8**, pp. 53-69.

Simmons, R.F. (1970) Natural language question answering systems: 1969. *Communications of the ACM* **13**, pp. 15-30.

Simon, H.A. (1969) *The Sciences of the Artificial* MIT Press, Boston, Massachusetts.

Smith, J.M.; Chang, Y.P. (1975) Optimizing the Performance of a Relational Algebra Database Interface. *Communications of the ACM* **18**(10), October, pp. 568-579.

Sparck-Jones, K. (1973) Index term weighting. *Information Storage and Retrieval* **9**, pp. 619-633.

Sparck-Jones, K. (1974) Automatic indexing: a state of the art review. *Journal of Documentation* **30**, pp. 393-432.

Stefik, M.; Conway, L. (1982) Towards the Principled Engineering of Knowledge. *The AI Magazine* **3**(3) Summer, pp. 4-16.

Thorne, J.P.; Bratley, P.; Dewar, H. (1968) The syntactic analysis of English by machine. *Machine Intelligence* **3**, pp. 281-297. Edinburgh University Press, Edinburgh.

Torsun, I.S. (1981) Dynamic analysis of Cobol programs. *Software Practice and Experience* **11**, pp. 949-961.
Toulmin, S. (1958) *The Uses of Argument.* Cambridge University Press, Cambridge (reprinted 1976).
Turing, A.M. (1936) On computable numbers, with an application to the entscheidungs problem. *Proceedings of the London Mathematics Society* **42** (2) pp. 230-265.
van Lawick-Goodall, J. (1971) *In the Shadow of Man* Collins, London.
Walker, D.E. (1975) The SRI speech understanding system. *IEEE Transactions on Acoustics, Speech and Signal Processing* ASPP-23, pp. 397-432.
Walter, G.W. (1953) *The Living Brain* Duckworth, London, also published by Pelican Books, 1963.
Wasen, P.C.; Johnson-Laird, P.N. (eds) (1968) *Thinking and Reasoning* Penguin Books, London.
Weber, H. (1976) A semantic model of integrity constraints on a relational data base. In *Modelling in Data Base Management Systems* (G.M. Nijssen, ed.), North-Holland, Amsterdam and New York.
Whitworth, W.A. (1901) *Choice and Chance* (5th edition), pp. 567-580 (reprinted by Stechert 1942).
Wilensky, R. (1984) *LISPcraft* W.W. Norton, New York.
Wilks, Y.A. (1972) *Grammar, Meaning and the Machine Analysis of Language* Routledge and Kegan Paul Ltd, London.
Wilks, Y.A. (1973) An artificial intelligence approach to machine translation. In *Computer Models of Thought and Language* (R.C. Schank and K.M. Colby, eds). W.H. Freeman, San Francisco.
Wilks, Y.A. (1975a) A preferential pattern-seeking semantics for natural language inference. *Artificial Intelligence* **6**, pp. 53-74.
Wilks, Y.A. (1975b) An intelligent analyser and understander of English. *Communications of the ACM* **18**(5).
Wilks, Y.A. (1977) Methodological questions about artificial intelligence: approaches to understanding natural language. *Journal of Pragmatics* **1**, pp. 69-84.
Winograd, T. (1972) *Understanding Natural Language* Edinburgh University Press, Edinburgh.
Wittgenstein, L. (1953) *Philosophical Investigations* (English Translation by G.E.M. Anscombe, Basil Blackwell, Oxford, 1974).
Woods, W.A. (1968) Procedural semantics for a question-answering machine. *AFIPS*, 33 (Part 1), pp. 457-471.
Woods, W.A. (1970) Transitional network grammars for natural language analysis. Communications of the ACM **13**, pp. 591-606.
Young, R.M. (1979) Production systems for modelling human cognition. In *Expert Systems in the Micro-electronic Age* (Proceedings of AISB Summer School) (D. Michie, ed.), pp. 35-45. Edinburgh University Press, Edinburgh.
Zadeh, L.A. (1965) Fuzzy sets. *Information and Control* **8**, pp. 338-353.
Zipf, G.K. (1949) *Human Behaviour and the Principle of Least Effort* Addison-Wesley, London.

Author index

Addis, T.R. 10-11, 50, 59, 113, 134, 158, 223-4, 227-8, 262
Ashby, R.W. 151-2, 296
Austin, J.L. 17

Babb, E. 10, 50, 102
Babbage, C. 13
Bandler, W. 31, 60
Banerjee, J. 106
Belady, L.A. 80
Bennet, J.S. 225
Bennett, J.M. 249
Bobrow, D.G. 287
Booth, A.D. 267
Bourbaki 31
Bramer, M. 12, 242
Buchanan, B.G. 223, 227
Bundy, A. 14

Carmichael, J.W. 12
Cendrowska, J. 242
Chang, Y.P. 87
Chapanis, A. 129
Charniak, E. 289
Chen, P.P. 48
Chomsky, N. 269, 288
Church, A. 194
Clark, D.W. 250
Codd, E.F. 27, 29, 115
Conway, L. 14
Coulouris, G. 11, 98

Date, C.J. 40, 43, 49, 82
Davis, R. 226
de Dombal, F.T. 229
Dreyfus, H.L. 153, 181
Duda, R. 224-5

Earl, L.L. 279, 281, 293
Edwards, W. 229
Efstathiou, J. 239
Elleby, P. 131

Fagin, R. 33
Fairthorne, R.A. 250
Feigenbaum, E.A. 223, 225
Fillmore, C.J. 276, 277
Fodor, J.A. 270, 277
Forrester, J.W. 47
Fox, J. 226, 232
Friedman J. 269
Fu, K.S. 224

Genrich, H.J. 295
George, F. 11
Glucksberg, S. 265
Goguen, J.A. 234
Goldman, N.M. 286
Gough, P.B. 181
Grant, T.J. 131
Green, L.E.S. 250
Greenwood, R.M. 250

Hall, P.A.E. 87
Hanna, F.K. 154, 271
Harrison, B. 16
Hartley, R.T. 11, 17, 25, 134, 228
Hayes, P.J. 24
Hendrix, G.G. 273
Hewitt, C. 284
Hinton, G.E. 296
Hodges, W. 188-90
Hubel, D.H. 152
Hudson, R.A. 283

Jackson, M.A. 154
Jaques, E. 49
Johnson, L. 10, 12, 15, 17-19, 139
Johnson-Laird, P.N. 153

Katz, J.J. 270, 277
Kay, M. 269, 270
Kellogg, C.H. 278
Kelly, K.J. 129
Kent, W. 296

Keravnou, E.T. 139
King, P. 10
Kohout, L.J. 12, 31, 60
Kowalski, R. 194, 204
Krauss, R.M. 265

Lakatos, I. 69, 263
Lautenbach, K. 295
Lefkovitz, D. 102
Lehman, M.M. 80
Lenat, D.B. 26
Lenneberg, E.H. 266
Lyons, J. 266

McCalla, G.I. 265
McCarthy, J. 93
McCleary, R.A. 152
McCorduck, P. 151
Maier, D. 33, 44
Maller, V. 10, 50
Mamdani, E.H. 239
Mandlebrot, B. 250-1
Maron, M.E. 278
Martin, M.V. 262
Meredith, P. 19
Michie, D. 225, 248
Minsky, M. 286-7
Mitchell, R.W. 11, 98
Moore, R.Y. 152
Mylopoulos, J. 265

Newell, A. 14, 26, 152, 181
Nilsson, N.J. 11, 158-9, 166, 169, 175, 176, 177-8, 217
Nordyke, R.A. 229

Oddy, R.N. 224

Pareto, V. 250
Pauker, S. 139, 226
Peirce 139
Peterson, J.L. 132, 295
Phillips, B. 219, 271
Pople, H.E. 226

Quillian, M.R. 271, 272
Quinlan, J.R. 227

Reade, C. 11
Rieger, C.J. 284
Ritchie, G.D. 154, 271
Robison, H.R. 279
Rogers, W. 93
Russell, B. 31

Sacerdoti, E.D. 227
Sampson, R.J. 265
Scarrott, G.G. 11, 249, 251-2, 289
Schank, R.C. 284, 288
Schröder 31
Shaket, E. 231, 234, 238
Shannon, C.E. 249
Sharman, G.C.H. 114
Shortliffe, E.H. 224-5, 245, 262
Simmons, R.F. 266, 277, 289-90
Simon, H.A. 16, 47, 152, 181
Smith, J.M. 87
Sparck-Jones, K. 227
Stefik, M. 14

Torsun, I.S. 80
Toulmin, S. 18, 240
Turing, A.M. 194

Walter, G.W. 151
Wasen, P.C. 153
Weaver, W. 249
Weber, H. 52
Whitworth, W.A. 250
Wiener, N. 31
Wiesel, T.N. 152
Wilensky, R. 197
Wilks, Y.A. 11, 274, 282, 288, 290-2
Winograd, T. 50, 234, 283, 288, 291, 293
Wittgenstein, L. 252
Woods, W.A. 269, 270

Young, R.M. 133

Zadeh, L.A. 231
Zipf, G.K. 249-52

Subject index

abduction 139, 222, 296
abstracting 278
abstraction, within organization 47-9
ABSTRIPS system 227
access paths 89
acquisition
  and elicitation 68
  of knowledge 226-9
action-symptom pairs 136
adjectives 236, 268-9, 272
address mode 110
adverbs 236, 272
afterset 60
amelioration 87
answer tree 202, 211
arcs (in graphs) 167, 168, 172, 175, 211
  labelled 271, 273
arguments 181, 186-9
  compound 186
  counterexample 190-1, 200
  fallacious 188
  field of 240
  force of 241-2
  valid 186, 187, 189-91, 222
arrays 158
artificial intelligence 13-14
  expectations of 13-14
  'hard' school approach 14, 154
  and human behaviour 249
  as reasoning process 182
  'soft' school approach 14
  symbol manipulation 16
  see also artificial intelligence programs
artificial intelligence programs 151-79
  decomposable production systems 174-9
  development 151-4
  graph search 167-74
  and human behaviour 181, 248
  production systems 154-66
Aspects of the Theory of Syntax (Chomsky) 269
assertions 239-42
associative law 197, 213
ATLAS computer 135
attributes 32
  expanding 75
  identifying 72
  key 52, 95, 120, 121, 123, 126
  linking 109, 110
  non-key 38
  OWN 32, 37, 64, 95, 120, 121
attribute sets 33
Automated Language Processing Advisory Committee (ALPAC) 268

Bachman diagrams 94
backtracking
  CRIB system 138, 139, 141
  control strategy 165-6, 172, 203, 214
backward systems 206-7, 216
bacteria 231, 242
Bayesian Theory of Classification 232-4
Bayes' rule 232, 234, 239
bijection 67, 68
Birkbeck College 267
bit maps 142, 143
Boyce/Codd normal form (BCNF) of tables 38-9, 45, 49, 67, 103
  intension 62
brain 151, 152
Brunel University 134
business systems 218

CAFS *see* Content Addressable File Store
calculus 182-93
  interpretation 185-6
  monotonic 187

predicate 182, 187, 191-4, 205
propositional 182-5, 187, 191, 194
cardinality 59-61, 77, 118, 129
card sorters 81, 93
Case for Case, The (Fillmore) 276
case grammar 276-8, 288, 289-90
case markers 279
certainty
criteria of 15
range of 243, 244
chaining 205
clauses 194-6
formation 195-6
goal 201, 202
Horn 204, 205
resolution 199
codomains 33
Common Business Oriented Language (COBOL) 27, 92, 93, 110
communication
man-computer 14, 113, 129, 228, 265, 283
in organizations 47-9
commutative law 197, 213
Computer-Based Medical Consultations: MYCIN (Shortliffe) 245
Computer-detectable Semantic Structures (Robison) 279
computer files *see* tables
Computer Retrieval Incidence Bank (CRIB) system 135-50, 224, 227-8, 230
assessment 145-6
on CAFS 140-6
codes 143
contact with distant groups 146-7
example of use 144-5
hierarchical structure 148
programs 135
Sfd graph 136
simplified system 147-9
computers
attitudes towards 24
generations 13-14, 24
and language 17
knowledge-based 14
see also computer understanding
computer stores 249
computer understanding 265-96
case grammar 276-8
first generation systems 267-8
frames and scripts 286-8
second generation systems 267, 268-70
semantics 270-6, 281-6
word government 278-81
concepts 19, 48
conceptual categories 284-6
conceptual database 79-111, 113-32
and conceptual model 58-9, 78
construction 80
consultant system 133-50
development 79-80
in ERA 53, 64
interpreter 113-32
as knowledge base 22-3, 25
logical models and schemas 81-97
in medical diagnosis 134
re-specification 80
conceptual model 47-78, 110-11, 113
effectiveness 128-9
ERA 62-77
extension 68, 131, 221
implementation techniques 243
intension 68, 131
interaction with language 128
progression of 72-7
query use 59
representations 58-62
semantic nature 78
task domain 219
conclusion (of argument) 186-7, 200, 204, 243, 244
Conference on Data Systems Languages (CODASYL) 27, 28, 92, 93
conjunction (Ands) 188, 192, 194, 207
of literals 215
connection trap 49, 94, 95
consistency of propositions 188, 194, 208, 214
checking 207, 212, 216
constants 191-7, 206
constraints
cardinality 59-61, 77, 118, 129
database 156-7
domain 129
between entities 52-4, 59
entity set 246
sfd 58-9, 61
transitive 123
update 59, 61, 77
consultant systems 133-50
CRIB on CAFS 140-6
CRIB system 135-40
simplified CRIB system 147-9
content accessing mode 110
Content Addressable File Store (CAFS) 49, 81, 113, 116, 120
architecture 100, 105
central processor 98-101
in CRIB system 135, 140-6
development 97-101

JNF data structures 102-4
MKI 98, 99, 101, 134, 140-2, 147
MKII (800) 98, 101, 106
searching 104-10, 121
storage structures 97-111
control information 206
context tree 228
control strategies 154-5, 163-6, 203-4
ancestry-filtered form 204
breadth-first 203
heuristic 155
irrevocable 163-5
linear-input form 203, 204
set-of-support 203, 204
tentative 165-6
unit-preference 203
conversations
acceptable 15
computerized 268
man-computer 129, 228, 265, 283
unacceptable 15
'cost' (in graph search) 168, 170-3, 175, 177
counterexample of argument 190-1, 200
count maps 142, 143, 147-8
count words 143
criteria 240-2
cybernetics 151, 152, 296

databases
constraints 156
design 27-30, 110-11
global 154-60, 167, 208, 216
'perfect' 30
relational 28
purchasing 87, 95-7
see also conceptual database
Data Description Language (DDL) 128
Dataflow Computer 81
Decision theory 230, 232, 239
decision tree 255-6
deduction 21, 139, 222, 296
deletion 125-8
demon rules 161, 288
DENDRAL system 223-7
denormalization 91
determinants 36-7
determiners 237
diagrams, as aid to explanation 19-20
dictionary 282, 291, 292
differencing 104-5
Direct Access Store (DAS) 98
discriminant function 234
disjunction (Ors) 188, 192, 194
of literals 207, 208, 211, 215
of rules 257
Disk Controller Unit (DCU) 98
Distributed Array Processor (DAP) 81
documents 19-20
dog and cat problem 207-17, 219-21
domains 31-2, 33
drugs 242, 243, 247

EDS 8 disk 99
element tables 35
elicitation 68-77, 218
and acquistion 68
progression of conceptual models 72-7
theory 69-72
EMYCIN system 224
entities
abstracted from environment 48-9
atomic 48
linguistic role 53
modelled in BCNF table 62
primary 48
properties 53
restrictions 65-6, 76
uniqueness 62
entity sets
ERA 219
extension 219
intension 219
normalized 91
relationships between 54-7, 74-5, 77
in storage structures 94, 95
table model 52, 53, 63
entropic decrement 255, 258, 261
entropy 252-6
communicable 252
usable 252
epistemics, applied 19-21, 222-63, 295
knowledge refining 249-63
MYCIN 242-7
types of epistemic systems 247-9
uncertain knowledge 229-42
see also under names of systems
evaluation function 163, 171
exchange problem 174, 176, 177
expert systems 18, 21-2, 139
epistemic 247-9
MYCIN model 242-7
range of 223-6
expressions (compound propositions) 183
Extended Relational Analysis (ERA) 17, 51-8, 102, 128
cardinality 59-61

in CRIB 137
development 130-1
effect of multivalued and join dependencies 62-8
as elicitation technique 68-77
ERA 1 131-3
ERA 2 131
ERA 1+ 131
introduction 51-3
limitations of 129-32
logic problem 218-21
in machine translation 266
reconstruction 56-8
and relational analysis 51
semantic functional dependencies 53-8

factorizing out 66
facts
conjoined 219
corroboratable 69-72
derivation 20-1
inexact 230
in rule-based systems 206-8
and theories 69-71
fault-finding 227-8, 252-3
fault reports 135, 227
fields
CAFS 100-1, 105
groups 101
header 103, 104
identifier 100
lengths 102
offsets 101
trailer 101
fifth normal form (5NF) of tables 44
File Correlation Unit (FCU) 109
file design 27-9, 58
files
'denormalized' 91-2
'flat' 102
'normalized' 30
of records 90
as relations 31
scanning 92
as tables 31
first generation systems 266-8, 277
first normal form (1NF) of tables 36-7, 73
Flexible Interrogation and Declaration Language (FIDL) 113, 114, 118, 299
Flexible Language Interpreter (FLIN) 113-32, 198
in CRIB system 135
Data Definition Language 114
examples of deletion 125-8
examples of insertion 121-4
mechanics of insertion 119-21
mechanism of implicit queries 115-19
system outline 114-15
focus imposition 257
force (of argument) 241-2
foreset 60
forwards systems 208-12, 213, 216
fourth normal form (4NF) of tables 40-2
frames 286-8, 289
'Frames paper' (Minsky) 286
functional dependencies 32-3, 36, 38
constraints 39-40, 53-4
on identifiers 73
on non-identifiers 74
transitive 38, 74
update problems 39
functions 191-2, 197
discriminant 234
fuzzy 239, 243, 244
Kasus 277, 279
loss 232-3
Skolem 195, 196, 208, 211
symbols 193
function words
change 115
create 115
delete 115, 128
erase 128
fuzzy logic 241
Fuzzy Semantic model 232, 234-9
Fuzzy Set theory 230-2, 239

General Problem Solver (GPS) 152
General Understander System (GUS) 287
generations of computers 13-14, 24
geological systems 224, 225
grammar 19, 183
context-dependent 269
context-free 268
systemic 283, 288
transformational 269, 288, 290
graph search 166-74
algorithm 'A' 168-73
task domain knowledge for 171-4
uninformed 170-1
graphs
AND/OR 175-9, 207-9, 212, 215
input 168
solution 209
see also graph search
group theory 151

hardware faults 134, 137
hedges (linguistic) 232, 236, 238
heuristics 21-3
  control strategies 155-6, 163, 172-3, 203
  in CRIB 138
  definition 21
hill climbing 163, 164
Horn clauses 204, 205
human factor 225-6
hyperarc 175
hypergraph 175
hypotheses
  auxiliary 71, 77
  context-dependent 208
  in CRIB system 139-40
  in Decision theory 232-4
  goal 206, 226, 244
  kinds of 20-1

IBM System R 82
image subset 33
implications 184-5, 204, 208, 211
indexing
  automatic 278-9
  inverted 32, 81, 98, 106, 107
  search area 106-8
  sequential 32, 37, 81, 98, 106
induction 140, 222, 227, 257-8, 262-3
industry 153-4
inference modes 21
inference rule systems 194-5
  complete 194
  refutation complete 194
infinitives 280
Information Processing Language (IPL) 268
information retrieval systems 133-5, 140, 198, 224
information storage
  insertion into database 115
  and knowledge representation 50
  removal from database 115
  structures 87-92, 92-7, 97-111
information theory 151, 152, 251
inheritance of properties 65, 66
input graphs 168
insertion
  examples of 121-4
  mechanics of 119-21
insight 153
intelligence 151, 152
INTERNIST system 226
interpretation 185-6
  unsatisfiable 215
interpreters 22, 113-32
In the Shadow of Man (van Lawick-Goodall) 289
Introduction to Database Systems (Date) 49
irrevocable strategies 163-5

join dependencies 42-5
  in conceptual model 76
  constraints 43, 64
  effects on ERA 62-4
  nature of 43
Joined Normal Form (JNF) 141
  data structures 102-4, 106
  tables 116, 121
joins 62-4
  hardware 108-10
  JNF 102
  multiway 64
  natural 34-5, 85, 141
  'non-loss' 35
  non-natural 84
  self-joins 109, 156-8
justification, and knowledge 17-19

Kasus functions 277, 279
k-connectors 175-7, 209, 215
key registers 98, 100, 102
keys
  candidate 38-9, 44
  composite 38
  overlapping 38
  primary 32, 36
  of tables 62-4
keywords 229, 268, 282
knowledge
  acquisition 226-9
  approach to 14-19
  asserted 207
  communication of 20
  of concepts 19
  consistency of 15
  control 206
  justification and 17-19
  level of 16-17
  manipulation of 20
  mini-theory of 24, 25
  storage of 20
  structural 206
  theory of 23-4
  uncertain 229-42
  see also knowledge refining, knowledge representation
knowledge base 20-3
knowledge elicitation techniques 16, 17, 25
knowledge engineer 25, 153, 295

knowledge engineering 23-6, 153
knowledge refining 249-63
    growing trees from entropy 252-6
    recursive 258-60
    Scarrott's conjecture 249-52, 262
    tree maintenance 260-2
    trees 256-8
knowledge representation
    and information storage 50
    manipulation 20
    schemes 79-80, 295
    storage 20
    techniques 25-6

Lakatos Programme 69, 140, 222
Lambda calculus 93
language
    interaction with conceptual model 128
    and meaning 16
    see also natural language
Law of Requisite Variety 152
legal extension 39
linguistics 266-74, 288, 296
list processing (LISP) data storage 93, 197, 250
lists 158
literals 191, 194, 205, 210, 215, 216
    antecedent 204
    conclusion 204
    disjunction 207, 208, 211
    goal 211
literature 19, 20
Lockheed Palo Alto Research Laboratory 279
logic 218-22
    limitations 221-2
    monotonic 18, 187, 296
    non-monotonic 187
logical connectives 116, 183, 184, 192, 209, 238
logical equivalence 186, 187
logical models 81-97
    hierarchical storage structures 87-92
    relational algebra 82-7
    ring structured storage 92-7
logical types 240
Logic Theory (LT) machine 152
loss function 232-3

machine reason 217-22
    ERA of logic problem 218-21
    limitations 217-18, 221-2
machines
    as epistemic mechanism 19
    general theory of 152
    intelligent 152
machine translation 266-75, 282, 291
    first generation systems 267-8
    second generation systems 268-70
MACSYMA system 225-6
magnetic tape 29, 81
management information system 22
map address 143
mapping function 155
mappings
    intensional 57
    of tuples 54
Markov chain 251-2, 260
mask patterns 101
mass processing 24
mathematics
    axioms 240
    methods of justification 18
    programs 225
meaning
    and knowledge 16
    and memory 281-6
    and understanding 15-16
measurement 238-9
mechanical reasoning 181-222
    calculus of reason 182-93
    limits 217-22
    reasoning with predicates 193-204
    rule-based systems 204-17
medical diagnostics 133-5, 204-5, 250
    MYCIN system 204, 224-31
membership functions 235-8
membership values 230-1, 237-9
memory
    conceptual 284, 286
    solid state 108-9
Meta-DENDRAL program 227
meta-reasoning 23
Methodological Falsification Programme 69
Modus Ponens argument form 187-8, 200
Modus Tollens argument form 187
monotone restriction 173, 177
morphemes 288, 293
most general unifier (mgu) 197, 213, 216
multihead reading 110
multivalued dependencies (MVD) 42, 44
    in conceptual model 76
    effects on ERA 62-4
MYCIN system 204, 224-9, 231, 242-7

natural language 265-96
    analysis 265

changes 291
English 265, 282
French 282
machine processing 265
in man-computer conversations 50, 129, 228, 231, 234
negative feedback 152, 162-3
networks 158
Newton's laws 23-4
NIL clauses 194, 200, 201, 203-4, 215
nodes in graphs 167-70
ancestor 168
conjunctive 188
connected 188
descendent 168
goal 170, 177
labelled 271, 272, 275
leaf 168, 209, 210, 211
root 168
start 177
supernodes 272-4
tip 168, 170
normalization 35-9, 49, 62
Boyce/Codd form 38-9
computer file model 35-9
fifth stage 44
first stage 36-7
fourth stage 40-2
relational model 39-45
restrictions 130
second stage 37-8
third stage 38-9
nouns 236, 237, 268-9, 280
numbers 237

1-connectors 175-6, 209, 215
optimum solution path 171-3, 177
organizations 47-9
output tree 168-70
OWN attributes 32, 37, 64, 95, 120, 121

paraplates 283, 288, 289
partitioning 273
partitions 66-8, 76-7
part records 29
pattern elements 98, 100, 102
pattern matching 196-9
pattern recognition 224, 229, 232, 239, 253
Petri nets 132, 295
physical model 89, 90, 95, 111
pointers
access 92
in logical models 81, 88-90, 93-5
'next' 93
power sets 36
Predicate Calculus 182, 187, 191-3, 194, 205
predicates 191-2
premises 243
Prenex form 196, 201
prepositions 269, 282
Present Illness Program (PIP) 139, 226
primitives 81, 271, 274
absolute 24
many-to-many 54, 57
many-to-one 54
one-to-one 54
semantic 282, 292
probability 238-9, 254, 260, 262
problem domains 229-30
problem solving 133, 152-3, 248
non-human 225
problem specification 14
production rules 161-3, 182, 203, 243
mandatory 161, 162
optional 162-3
production systems 154-66
AND/OR graphs 175-9
control strategies 163-6
decomposable 174-9
global database 156-60
production rules 161-3
three elements 154-6
program specification 14
projection 33-4
proof systems 194-5
Propositional Calculus 182-5, 187, 191, 194
propositions 182-3, 188-9
atomic 188, 189
compound 188, 189
consistency 188-9
True or False 182-6, 190, 191, 193
prose 19
PROSPECTOR system 224-6, 228
protocol analysis 181
Protosynthese 1 program 277
Pseudo Retrieval Language (PRL) 198
PSYCO system 226
purchasing system
key attributes 120-1
Sfd graph 118
specification 117
tables 120-1

quantification 192-3
existential 192, 211
universal 192, 211
quantifiers 211
queries 91, 110, 115, 135

implicit 115-19
question answering systems 50, 266-7, 269, 292
quorum function 100

RAFFLES system (Rapid Action Fault Finding Library Enquiry System) 227-8, 230, 253, 262
Rand Corporation 278
random access 29
random distribution 250
range of subset 33
real-time interaction 24, 25
reasoning
in computers 153, 225-6
deductive 218
goal directed 226
inexact 230
non-monotonic 18, 296
see also machine reason, mechanical reasoning
records
CAFS 100-3, 141
format 101-2
'linker' 95
in logical models 81, 88, 93, 95
'next' 93
ordering 105
recursive techniques 251-2
reductio ad absurdum 191, 200
refutation tree 201, 202
relational algebra 29-30, 35, 130
in logical models 82-7
operations 33-4, 39, 49
relational analysis 27-46
as design methodology 28
extensions 50
functional dependency 32-3
normalization 33-45
origins and development 27-30
role of 45-6
tables 30-2
and trans-relational analysis 49
relational operations 220
in mechanical analysis 278
natural join 34-5, 199
presentation 82
projection 33-4
role of 45-6
rules 161
relations
extensional 31
intensional 31
and tables 30-2, 34, 53
relationships
between entity sets 74-6, 88
many-to-many 76, 88
many-to-one 88
one-to-one 88
'remembered hits' 105
repeating groups 29, 35
in records 88, 90
resolution inference 196, 200
Resolution Principle 154, 220
resolution refutation 199-203, 206, 208, 211, 214, 215
retriever 117, 118, 125, 198
reverse element-set 60
right to be sure 17-18
ring structure 93
Risch algorithm 225
robot systems 227
root rule 116, 127
root tables 115-19
rule-based systems 204-17, 220, 257
achieving goals 211-12, 226
backward rules 213, 214-17
forward system 208-12
methods for 206-7
nature of rules 205-6
simple scheme 204-5
unifying composition 212-14
rule form 205-6
rule independence 18
rule induction 257-8
rules 245-7
backward 214-17
collection 227
constraint 247
data collecting 246
forward 210
leaf 212
nature of 205-6
structural 21
transformational 21
see also rule-based systems

satisfaction set 30-2
schemas 81
see also logical models
scientific programmes
non-progressive 71
progressive 70
scripts 286-8
search
AND/OR graph 177-9
breadth-first 170, 172
depth-bound 166
depth-first 140, 141, 170, 172, 173
differencing 104-5
graph 166-74

hardware joins 108-10
improving 104-10
indexing 106-8
task domain knowledge for 171-4
uninformed 170-1
variable trailers 105-6
search evaluation unit (SEU)
CAFS 98, 100, 102
second generation systems 267, 268-70, 279, 288
second normal form (2NF) of tables 37-8, 74
selector 116-18, 125, 198
semantic event form 278
semantic functional dependency (sfd) 53-8
constraints 58-9, 130
equivalence 56
reconstruction 56-8
0-implies 55, 65, 88, 91
1-implies 55-6, 63-5, 68, 88, 91, 120
see also sfd graphs
semantic nets 228, 271, 273-6, 278, 289-90
semantics 267, 270-6
preference 290
structures 270-2, 281-6, 293
and syntax 279, 290
sentences 15, 181-2, 266
analysis 283
declarative 182
embedded 277
generation 277
modality 277
proposition 277
syntax 269, 276
servant rules 161, 288
set intersection 108-9
set operations 130, 131
sets 93, 95
sfd graphs 102, 114, 289
directed graph 275-6
dog and cat problem 220
elicitation 68, 69, 72
joins 64, 65
machine-shop 85-6
MYCIN 242, 243, 245, 246
purchasing database 58-9, 87-8, 91-2, 95, 118-19
rule dependency 245, 246
student-course environment 77, 78
tiles problem 159, 161
town problem 156-7
updates 115
SHRDLU system 234, 283
signs 182
Skolem constant 195, 196, 201
Skolem functions 195, 196, 208, 211
software 17
faults 134
standards 240-2
Stanford University 282
state-goal function 163-5
statements 18
state transformations 151-2
statistics 229
storage elements 89
storage structures
CAFS 97-111
hierarchical 87-92, 106
linear 88
ring-structured 92-7
Structure of a Semantic Theory (Katz and Fodor) 270
sub-schemas 81
see also logical models
substitution 196-7
suckets 107, 108
supernodes 272-4
symbol manipulation 16
symbol processing 152
symbols 182
relational algebra 82, 88, 89-90, 94, 161
symbol strings 152, 196
symptom groups 136-8, 141-3, 147-8
key groups 136-8, 142, 143
subgroups 136-8, 142-5
symptom patterns 135
syntax 15, 267, 269
analysis 282
and semantics 279, 290-2
system development 14
systems analysis 27
Systems Development Corporation 278

tableau technique 189-91, 196, 200
tables
as aid to explanation 19
BCNF 39-42, 52, 58, 62, 67
column headings 31
CRIB 141-2
extension 32
intension 32
keys 62
normalization stages 36-45
and relations 30-2, 34, 53
union-compatible 82, 83
unnormalized 35-6
update restrictions 61-2

tape-based systems 94
task analysis 274-5
task domains 48, 52, 53
 analysis 50, 130
 business systems 218
 conceptual model 219
 industrial 153
 for search 171-4, 242
tautology 21, 186, 193, 215, 222
Teachable Language Comprehender (TLC) 271
teaching aids 22
technologist, role of 17
technology
 in communications 47
 in problem-solving 49
TEIRESIAS system 226-7
templates 282-3, 287-90
tentative strategies 165-6
termination condition 170
tests 252-60
theories
 initial 69
 non-progressive 71
 progressive 70
third normal form (3NF) of tables 38-9, 74
thought 153, 247, 249
threshold values 237
tiles problem 158-62, 164-6, 167, 174
town problem 155-7, 165, 167
traces (sets of symptoms) 253, 255
trailers, variable 105-6
training set 255-6, 262
trans-entity dependencies 53, 54
transformation rules 154, 155, 219-20
transition networks 270
trans-relational analysis 49-51
trans-relational dependencies 49
tree structures 93, 188-9
 answer trees 202, 211
 binary 258
 entropy 252-8
 generate 258-60
 maintenance 260-2
True or False propositions 182-6, 190, 191, 193
truth assignment 239
truth maintaining 131
truth selection 131
truth tables 183-5, 187
truth testing 131

uncertainty 239
understanding
 by computers 14
 and meaning 15-16
unifiers 197, 213
unification 196-8, 199, 207, 211-16
 collection 212-13
 composition 212-14
 substitutions 214
United States Census Bureau 93
universal relation 39-40
 blanks 40, 55, 56, 103
 constraints 53
 unknown values 40, 55, 56
universal specialization 194-5
universes 192, 266
 infinity of 193-5
update problems 30, 91, 146
updates 115, 242, 243, 248
users 113-15, 129, 149
Uses of Argument (Toulmin) 240

variables 191-8, 206, 208, 209
 bound 192
 free 192
vectors 103-4, 107
 fuzzy 238
 membership 116, 121
 'no additions' 128
 'no deletions' 128
 table-membership 103, 141
verbs 277, 279-81
visual display units (VDU) 15
von Neumann machine 81, 88

well-formed formulae (wff) 183-6, 208, 210
 answer 211
 goal 206, 211-12
 interpretation 185
 invalid 186
 resolution 199
 unsatisfiable 194
 valid 186, 194
Whitworth curve 250
word government 267, 278-81, 288-91, 293
words
 machine translation 269
 meaning of 16
 order 290
 in sentences 279, 290

Zipf's law 249-52